AF327088

COMPOSITE STRUCTURES

SAFETY MANAGEMENT

COMPOSITE STRUCTURES

SAFETY MANAGEMENT

Second Edition

B.F. BACKMAN

ELSEVIER

Amsterdam • Boston • Heidelberg • London • New York • Oxford
Paris • San Diego • San Francisco • Singapore • Sydney • Tokyo

Elsevier
The Boulevard, Langford Lane, Kidlington, Oxford OX5 1GB, UK
Radarweg 29, PO Box 211, 1000 AE Amsterdam, The Netherlands

First edition 2005
Second edition 2008

British Library Cataloguing in Publication Data
A Catalogue record for this book is available from the British Library

Library of congress Cataloging-in-Publication Data
A Catalog record for this book is available from the Library of Congress

ISBN: 978-0-08-054809-8

For Information on all Elsevier publications
visit our web site at books.elsevier.com

Typeset by Charon Tec Ltd (A Macmillan Company), Chennai, India
www.charontec.com

Prined and bounded in Hungary

08 09 10 10 9 8 7 6 5 4 3 2 1

Working together to grow
libraries in developing countries

www.elsevier.com | www.bookaid.org | www.sabre.org

ELSEVIER   **BOOK AID** International   Sabre Foundation

CONTENTS

Preface

This book is intended for composite safety specialists in industry, government and academia and for advanced graduate students with an interest in bettering their understanding of the challenges of safety in designing composite structure.

The steady stream of new materials, processes and new structural concepts has made empirical structural design approaches obsolete and the absence of pertinent service experience, especially with regard to transport-category airplanes, has forced a rethinking of the role of safety in structural design.

Innovation is the state of the evolution of composite structures. Explicit safety measures have to be introduced or innovation will not be manageable from a safety standpoint. Innovation comes with several levels of uncertainty and risk management, and monitoring of safety levels and control processes for safety level correction in service.

A well-defined 'family' of safety-specific 'elements of safety' will blaze the trail for future regulations in composites design, manufacturing, maintenance, operation and process development. Safety is a time-dependent variable. Requirements must be kept current and well-defined, while the means of compliance must be adaptable.

This book sets the stage for safety-based airplane structural development that takes advantage of innovation to improve airplanes and to make them safer and cheaper. It discusses different safety goals promoted by the Federal Aviation Administration and demonstrates 'what it takes.' The future of structures belongs to composites but only if they are introduced through safe innovation and use of explicit safety measures.

Part I
Safety Management Analysis

Chapter 1
Introduction

'Safety Management', a term used in regulatory circles during the past 10 years or so, needs to be supported by a detailed set of developments and design criteria in order to promote safe innovation for composites. The concept places its focus on the feedback obligations of most of the disciplines of the organizational 'elements of structural safety'. A successful introduction to the certification procedures will require detailed identification of safety issues in the initial phase and extensive rules for risk management, upgrading and updating, and associated validation of safety control. Backman (2005) discusses the elements of structural safety. The latter part of this book identifies the nature of regulatory activities from the standpoint of structural safety. The value of regulatory involvement is highlighted. Moreover, the book provides a detailed investigation of all the elements of structural safety and the contributions they make.

The introduction of Risk Management and Uncertainty Control into the process of achieving and maintaining safety levels necessitates a disciplined approach that recognizes that safety is a function of time. Any successful introduction of 'safe innovation' requires an activity that starts during initial development and continues until the end of service life.

The starting point for healthy Safety Assurance and Control implementation must be the realization that structural efficiency and safety do not evolve from opposing objectives. The process has to assure a rational definition of data and knowledge requirements at the start of the design process – the a priori insight, e.g. see Tribus (1969).

It also should embody regulation and compliance requirements for data acquisition and development of further knowledge during service. The monitoring process and the risk management process will produce an ever-increasing insight into existing levels of safety, and the actions required to produce the correct adjustments are essential and must be regulated. A useful basis is described in Congdon (2003).

The needs of explicit safety requirements for development of the processes of design, manufacturing, maintenance, operation and requirements are tightly linked to change and innovation. The challenges derived from new materials, new processes and new structural concepts are inherent to the use of composites for flight vehicles. The call for more efficient structures and safer aircraft rings more loudly now than ever. It is true that it is the responsibility of each individual discipline to implement processes that ensure the meeting of safety objectives. The failure to do so, for the most part results in structural integrity loss, so a coordinated set of requirements must be put in place, and monitored and enforced.

This book extends the role of Safety Management to include many more functions than initially was visualized. In addition to a feedback system that makes it possible to obtain a true description of the service environment over time, it is also desirable to include a regulatory mandate to cover both compliance requirements for the initial design phase and continual upgrading during service. A rigorous reporting requirement also should be implemented for all elements of structural safety. The assessment of the initial effects on safety, the correction of safety levels through risk management, and the development of regulations for 'designed-in' safety levels of emerging aircraft, derivatives and new models must be an integral part of a safety process that ultimately would produce new requirements for all elements of safety.

One of the most important aspects of this approach is the definition of the service environment, the initial identification of the extent of damage and probabilities of detection, types of damage, scale of external and internal damage, and residual strength. Errors in quality control, process failures, inspection irregularities, repair mistakes, and violations of operational limits must all come with an initial understanding of the scope of the requirements for dealing with design data and safety threats. The process must deal with the identification of the minimal levels of knowledge required to establish basic design data and knowledge bases. It must determine a basic compliance definition for roll-out, and a foundation for further safety level baselines derived by risk mitigation and reduction of uncertainty. It is not surprising that a required knowledge basis for successful design is a tangible quantity that can be defined and verified by monitoring processes and risk management, and it needs to be subject to regulations and compliance requirements in all its phases.

1.1 ELEMENTS OF STRUCTURAL SAFETY

Backman (2005) describes the interaction of the 'elements of safety', but focuses on the design element. The general definition of the elements of safety can be found through the examination of the following equation:

$$P(S_T) = P(S_D S_M S_I S_O S_R) \qquad (1.1)$$

This equation states that

the probability of a safe structure (S_T) is equal to the joint probability of safe design (S_D), safe manufacturing (S_M), safe maintenance (S_I), safe operation (S_O) and safe regulations (S_R).

Equation (1.1) can be expanded in the following manner when requirements, including their status, are given the roles they deserve. One version of Equation (1.1) is

$$P(S_T) = P(S_D \mid S_R) \cdot P(S_M \mid S_R) \cdot P(S_I \mid S_R) \cdot P(S_O \mid S_R) \cdot P(S_R) \qquad (1.2)$$

where the importance of requirements formulation has been highlighted.

The above equation can be used to produce good approximations for the probability of an 'unsafe structure'. The result is, for example, as follows:

$$P(\bar{S}_T) = P(\bar{S}_D \mid S_R) + P(\bar{S}_M \mid S_R) + P(\bar{S}_I \mid S_R) + P(\bar{S}_O \mid S_R) + P(\bar{S}_R) \qquad (1.3)$$

A study of Equation (1.3) reveals that contributions to the unsafe state by the five terms would be most effective if all were of equivalent magnitude. The first term was the focus of Backman (2005).

The second term represents the 'probability of unsafe manufacturing, given safe maintenance, safe operation and safe requirements'. The third term describes the 'probability of unsafe maintenance, given safe operation and safe requirements'. The fourth term is the 'probability of unsafe operation, given safe requirements'. The fifth term represents the 'probability of unsafe requirements'. It also represents the importance for regulators to accept a distributed, active role in expansion of Safety Management, an expanded responsibility that encompasses initial and continuing roles to support efficiency and safety in the introduction of innovation and support of composite structures in service.

Equation (1.3) identifies the primary elements of safety as:

- design
- manufacturing
- maintenance
- operation
- requirements formulation.

The next section contains a study of the latter four elements, the first, as mentioned above, being the subject of Backman (2005). Damage size plays an important role; this is illustrated in Figure 1.1.

Each of the elements can be split in to separate functions; e.g. design process (D_P) and process implementation (D_E).

Safe design can be factored as

$$S_D = S_{DD} \cdot S_{DC}$$

(Safe, design, drawing process (S_{DD}), and safe, design, compliance, demonstration (D_{DC})).

Safe manufacturing becomes

$$S_M = S_{MM} \cdot S_{MP} \cdot S_{MI} \cdot S_{MA} \cdot S_{MF}$$

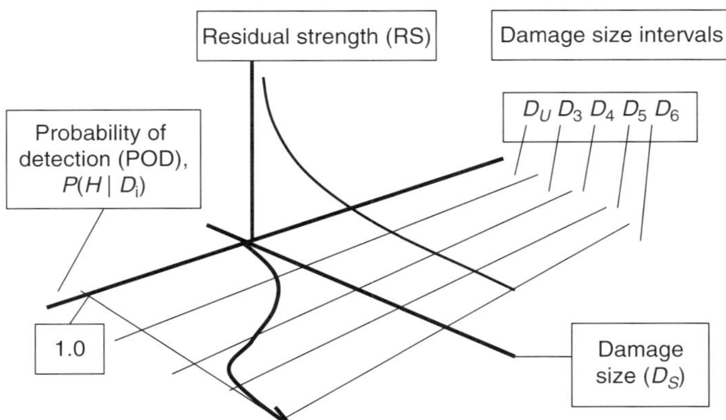

Figure 1.1 Damage size interval.

The first factor represents safe material processing; the second, processing; the third, safe installation; the fourth, safe assembly; the fifth, safe final assembly.

A common thread through all the elements of safety is the avoidance of unsafe states and, consequently, failures. The following equation describes the probabilities of failure after reaching an unsafe state:

$$P(\overline{A}) = P(\overline{S}_T \overline{A}) = P(\overline{U}_T \overline{H}_T \overline{A}) = P(\overline{A} \mid \overline{UH}_T) \cdot P(\overline{H} \mid \overline{U}_T) \cdot P(\overline{U}_T) \qquad (1.4)$$

This equation, if it were to describe a situation that makes 'walk-around' inspections feasible, involves the following events:

$$\overline{A} \; : \; \text{failure}$$
$$\overline{S}_T \; : \; \text{unsafe state reached at '}T\text{'}$$
$$\overline{U}_T \; : \; \text{loss of structural integrity at '}T\text{'}$$
$$\overline{H}_T \; : \; \text{damage not detected at '}T\text{', scheduled inspection}$$

$$P(\overline{A}_n \mid \overline{U}_T \overline{H}_T) = [1 - (p_D + \overline{p}_D p_S)^n]$$

where: p_D : probability of detection during 'walk-around'

p_S : probability of survival during a flight

n : number of flights.

The following example, (Example 1.1) describes how the probability of failure varies with n, the number of flights after undetected loss of integrity.

Example 1.1   We assume 'walk-around' capability and a normal distribution for both strength and internal loads. We also assume the values below for p_D and p_S, the probability of failure (lower limit for unsafe state), before the nth flight is

$$P(\overline{A}_n) = [1 - (0.6 + 0.4 \cdot 0.995)^n] \cdot 10^{-5} \cdot 10^{-6}$$
$$= (1 - 0.998^n) \cdot 10^{-11}$$

N	$P(\overline{A}_n)$	
1	$10^{-11} \cdot 0.002 = 2.0 \cdot 10^{-12}$	
1000	$10^{-11} \cdot 0.860 = 0.9 \cdot 10^{-11}$	
3000	$10^{-11} \cdot 0.997 = {\sim}10^{-11}$	◄ Upper limit in risk management

This table could be considered to deal with growth to the unsafe state, or accidental damage just after a major inspection. It could also be seen as an illustration of the fundamentals of risk management.

1.2   INTERACTION BETWEEN ELEMENTS

A review of the elements reveals that all deal with both different and similar aspects of safety:

- designed-in safety levels;
- built-in safety levels;
- maintained or restored safety levels;
- protected and reported safety levels;
- legislated and enforced safety levels.

This list corresponds to the elements of safety identified in the previous section. These are primary elements and an integral part of structural safety. There are of course interactions with other disciplines which affect the general safety situation, e.g. flight systems. These interactions can be dealt with in a way similar to that used for the elements of safety described here, but fall in the discipline of airworthiness.

Structural integrity and high-quality processes are the two main safety concerns for all the elements of safety; the common defects are:

- property reduction (e.g. strength);
- violations of maximum load limits (e.g. 'built-in' loads);
- aggravated damage (e.g. size).

The main objective of safety management is to control the formulation of safe process requirements and safe application of those requirements (control of defects produced), and, in relation to composites, to produce safe structures even in the scenario of innovation.

1.2.1 *Manufacturing and safety*

Manufacturing, specifically quality assurance and quality control, is a very important aspect of structural safety. Critical manufacturing elements are controlled by process and quality control.

One version of the term that measures the probability of unsafe manufacturing in Equation (1.3) is

$$P(\overline{M}) = P(\overline{S}_M \mid S_R) \tag{1.5}$$

Equation (1.5) shows the probability of interest, and the conditional probability of unsafe manufacturing, given safe requirements, which is particularly important for innovation.

The main threats are process specification violations that cause:

- low, structural properties;
- spurious local loads (e.g. clamp-up at bolt locations);
- aberrant geometry (e.g. incorrectly edged margins);
- unacceptable flaws.

All of these constitute threats to structural integrity. However unacceptable, manufacturing flaws are part of damage tolerance considerations and of course subject to feedback and updates, but can be a separate threat like the others (e.g. by compromising damage resistance and damage growth rates). Low, structural properties represent a direct threat to structural integrity, if specifically outside the limits of the manufacturing processes. Aberrant geometry can also cause a direct threat to local design data. Finally, spurious, local loads can increase design loads at specific points and their occurrence must be rendered very improbable by meticulous quality control.

The main interaction is with damage-tolerant designs. Manufacturing flaws (e.g. bond-line flaws), often have minor external damage (see Figure 1.2), in the region between 'a' and zero. Therefore, difficultly of detection will make the loss of fail-safe integrity very likely. The most appealing solution seems to be one that can make the occurrence of flaws very unlikely by means of thorough process control. There of course is interaction with inspection technology development, so, for example, detection that is more dependent on internal than external damage would be favored as a better choice.

Figure 1.2 shows a desirable trend in the focus on detection. A desirable situation would, in detection criteria, make internal damage more important than external damage. The interaction takes place between this situation and reducing the probability of occurrence

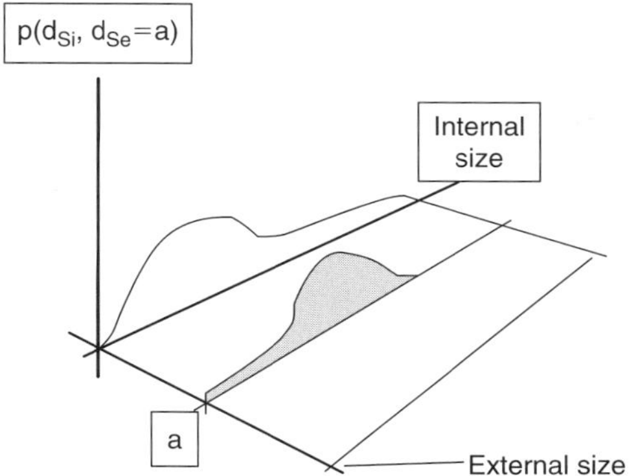

Figure 1.2 Damage relations.

of these types of damage. The inadequacy of the 'Tap-test' for thick gages in the structure is well documented, and the true choice is between new technology for detection or better process control reducing the probability of occurrence of this awkward flaw type. Also, it is the case that often when the properties are degraded, damage growth rates are increased, thus requiring more service inspections.

Loss of structural integrity, specifically damage tolerance integrity due to manufacturing process violations, can be expressed in terms of probabilities as

$$P(\bar{U}_L\bar{M}) = P[\bar{B}_L\bar{X}D_5\bar{X}_M(T_{Mi} \cup \cdots \cup T_{Mn})] = \sum_{i=1}^{n} P[\bar{B}_L\bar{X}D_5\bar{X}_MT_{Mi}] \qquad (1.6)$$

Equation (1.6) contains the following events:

$\bar{B}_L$: residual strengths $\leq LLR$
$\bar{X}$: mechanical damage is present
D_5 : limit damage size range
$\bar{X}_M$: manufacturing process failure
T_M : failure type and extent
$\bar{U}_L$: lost limit integrity.

A typical term on the right-hand side of Equation (1.6) can be written as

$$P(\bar{B}_L\bar{X}D_5\bar{X}_MT_{Mi}) = P(\bar{B}_L \mid \bar{X}D_5\bar{X}_MT_{Mi}) \cdot P(D_5 \mid \overline{XX}_MT_{Mi})$$
$$\cdot P(\bar{X} \mid \bar{X}_MT_{Mi}) \cdot P(T_{Mi} \mid \bar{X}_M) \cdot P(\bar{X}_M) \qquad (1.7)$$

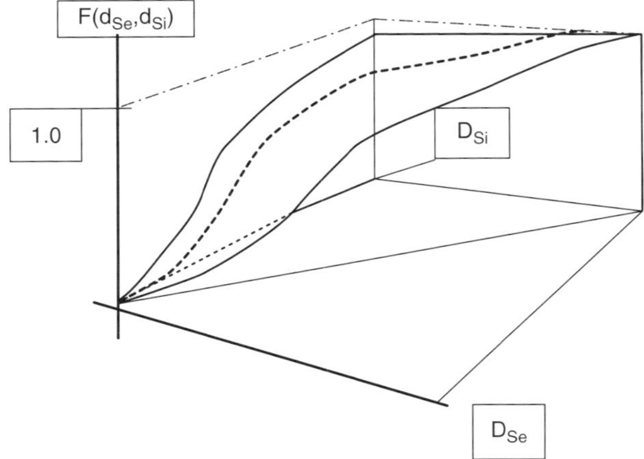

Figure 1.3 Detection probabilities; distribution function.

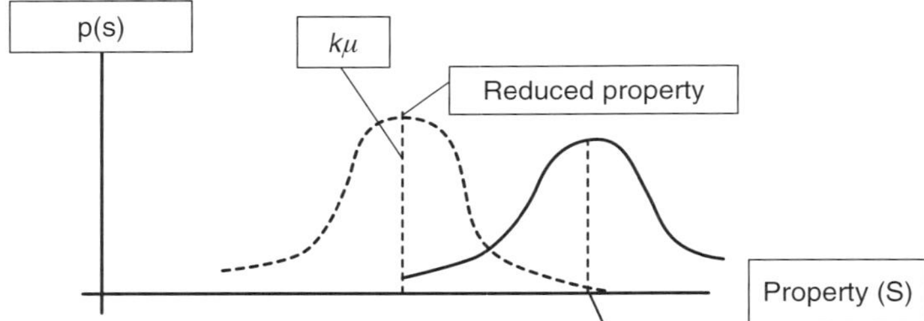

Figure 1.4 Example of property reduction.

Example 1.2 Figure 1.3 illustrates the expectations of order of magnitudes to achieve equivalent values for different elements. We now assume that the types have an equal probability of occurring and that the principal structural element (PSE), in question has 50 design points, each one with a design requirement of $2 \cdot 10^{-9}$ for the PSE. We also assume that the process failure causes a 10% reduction in a structural property.

We assume that 'k' in Figure 1.4 is 0.90 of the B-value for the 'pristine' material, which is 0.87μ. The B-value for the degraded material/structure has the following probability value in the 'new' distribution. With normally distributed properties we get

$$\Phi\left(\frac{0.87\mu - 0.9\mu}{C_V\mu}\right) = \Phi(-0.3) = 0.38$$

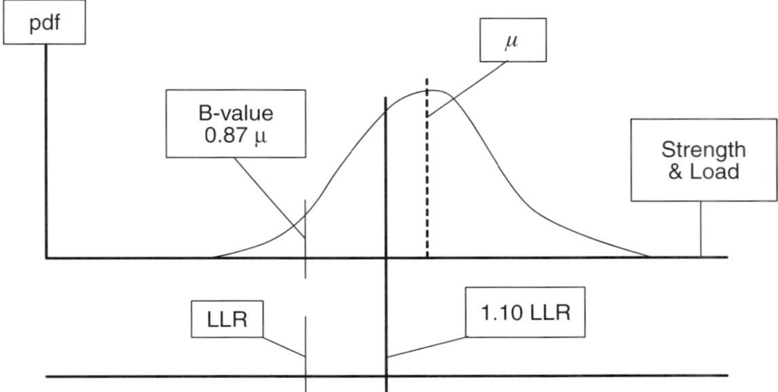

Figure 1.5 Change in design load.

For this case the equation would yield a probability of

$$0.4 \cdot 10^{-3} \cdot 10^{-2} \cdot 1 \cdot 10^{-5} = 4 \cdot 10^{-9}$$

for each design point.

The interesting observation is that for a requirement for design, deemed reasonable in Backman (2005), we would expect the probability of process failure at a design point to be, at most, 10^{-5}, in order to be equivalent to what the basic requirement for limit design integrity is.

Mishaps in manufacturing can also cause local loads that threaten the integrity of the structure. Figure 1.5 illustrates the effect.

We assume a 10% increase and normally distributed strength, the resulting probability being

$$\Phi\left(\frac{1.10 \cdot 0.87 \cdot \mu - \mu}{C_V \mu}\right) = \Phi(-0.43) = 0.33$$

The probability of violating the residual strength integrity is 33%.

Manufacturing mishaps that result in geometry aberrations can result in a strength reduction, (e.g. edge margin reductions). Assuming an error in geometry that causes a 10% reduction in strength, the following probability results:

$$\Phi\left(\frac{0.87\mu - 0.90\mu}{C_V \mu}\right) = \Phi(-0.30) = 0.38$$

Finally, manufacturing mishaps that cause damage beyond imperfections allowed by the process specifications raise the possibility of accidental, large-scale damage in the vicinity of a manufacturing-induced flaw. The probability of limit integrity is lost under these circumstances, and can be written as

$$\begin{aligned}
P(\bar{U}_L \bar{X}_M T_{M4}) &= P(\bar{B}_L \mid \bar{X}_M T_{M4} \bar{X} D_5) \cdot P(D_5 \mid \overline{XX}_M T_{M4}) \cdot P(T_{M4} \mid \bar{X}_M) \\
&\quad \cdot P(\bar{X}_M) \cdot P(\bar{X}) \\
&= 0.33 \cdot 10^{-3} \cdot 0.25 \cdot P(\bar{X}_M) \cdot 10^{-2} = 0.8 \cdot 10^{-6} \cdot P(\bar{X}_M)
\end{aligned}$$

So in order to produce 'an equivalent order of magnitude', the following inequality must be satisfied:

$$P(\bar{X}_M) \le 10^{-4}$$

The probability of causing loss of limit integrity due to a 'manufacturing flaw', therefore, should satisfy the inequality

$$P(\bar{U}_L \bar{X}_M T_{M4}) \le 0.8 \cdot 10^{-10}$$

which then in total is

$$P(\bar{U}_L \bar{X}_M) = 3.55 \cdot 10^{-10}$$

1.2.2 Safety and maintenance

The role of unsafe maintenance can be defined by the third term of the right-hand side of Equation (1.3):

$$P(\bar{I}) = P(\bar{S}_I \mid S_R) \tag{1.8}$$

Maintenance, periodic scheduled maintenance, inspection and repairs, come with the following types of defect:

- induced damage
- added secondary loads
- inaccurate inspections
- inadequate quality of repair.

The threat to limit integrity from 'unsafe maintenance' can be expressed as

$$P(\bar{U}_L \bar{I}) = P(\bar{B}_L \bar{X} D_5 \bar{I} (T_{I1} \cup \cdots T_{In})) = \sum_{i=1}^{n} P(\bar{B}_L \bar{X} D_5 \bar{I} T_{Ii}) \tag{1.9}$$

Here T_{Ii} represents the type and extent of the threats.

'Induced damage' due to 'unsafe maintenance' tends to be caused either by careless removal of parts or sub-assemblies or by faulty re-installation. However, it may also have its roots in inadequate scheduled maintenance resulting in 'excessive' wear and tear.

The introduction of 'added secondary loads' may also have the above causes, but, in addition, it can be traced back to imprudent implementation or design of repairs.

Inadequate inspections result in damage being present longer in the size range constituting a safety threat to what is prescribed in the design criteria. This threat is another reason for prescribing a ubiquitous, fail-safe design of all PSEs. The main effect here is a loss of ultimate integrity, which clearly compromises fail-safe integrity.

Finally, inadequate quality of repair, interpreted as not successfully restoring ultimate strength, can be detrimental to future required ultimate integrity in support of fail-safe integrity and is therefore a safety threat. The picture of interaction between different kinds of anomaly in manufacturing and maintenance procedures is beginning to emerge in terms of both interaction and effects.

SUMMARY

Induced damage has, in principle, the same consequences as the manufacturing flaws, and threat probability can be compiled in the same manner. It mainly constitutes a threat to ultimate integrity and its importance to fail-safety. It also involves more than isolated damage, which must be evaluated from a criticality standpoint in the design of damage-tolerant structures.

Added secondary loads due to maintenance mistakes have effects very similar to those resulting from manufacturing errors, and could be dealt with in the same manner.

Inaccurate inspections and *inadequate repair quality* both constitute threats to fail-safe integrity but are also the source of interaction with manufacturing mishaps.

Example 1.3   The main objective of this example is to investigate the interaction with manufacturing flaws. If we accept the assumption that 'inadequate repair quality' creates a similar situation to manufacturing flaws then the combination of the two results yields the following probability:

$$P(\bar{U}_L\bar{S}_C) = P(\bar{B}_L \mid \bar{X}D_5\bar{S}_C) \cdot P(D_5 \mid \overline{X S_C}) \cdot P(\bar{X} \mid \bar{S}_C) \mid P(\bar{S}_C)$$

where $\bar{S}_C = \bar{X}_M T_{Mi}\bar{S}_{Ii}$

The use of the same type of data as in Example 1.3 results in the following order of magnitude:

$$P(\bar{U}_L\bar{S}_C) = 0.4 \cdot 10^{-3} \cdot 10^{-2} \cdot (10^{-4})^2 = 0.4 \cdot 10^{-13}$$

which indicates that the combination of these kinds of event may turn out to be very unlikely. The focus seems to be best directed toward the 'simple' events.

1.2.3 Safety and operation

The third term in Equation (1.3) deals with the probability contribution of operation to an unsafe state:

$$P(\bar{O}) = P(\bar{S}_O \mid S_R) \tag{1.10}$$

The threats to safe operation derive from:

- exceeding limit load (outside placard);
- exposing the structure to unsafe environments (e.g. keeping the structure for several hours at maximum or minimum temperature);
- not reporting 'ground damage' promptly.

The probability that operational mistakes will cause a loss of limit integrity can be written as

$$P(\bar{U}_L\bar{O}) = P(\bar{B}_L\bar{X}D_5\bar{O}(T_{O1} \cup T_{O2} \cup T_{O3})) = \sum_{i=1}^{3} P(\bar{B}_L\bar{X}D_5\bar{O}T_{Oi})$$
$$= \sum P(\bar{B}_L \mid \bar{X}D_5\bar{O}T_{Oi}) \cdot P(D_5 \mid \overline{XOT_{Oi}}) \cdot P(\bar{X} \mid \bar{O}T_{Oi}) \cdot P(T_{Oi} \mid \bar{O}) \cdot P(\bar{O}) \tag{1.11}$$

Here T_{Oi} refers to the types and extent of threat identified above.

The threat associated with exceeding limit load is a temporary anomaly, compared with most of the other threats discussed. However, if the excess causes permanent damage, then again the result will constitute a threat until detected.

Excessive environmental conditions can result in many different effects, secondary stresses causing a violation of the limit cut-off or causing damage. In hybrid structures (e.g. titanium and composites), permanent 'built-in' stresses and damage can result and the situation has to be dealt with on a 'case-by-case' basis.

Unreported 'ground' damage can leave the structure with damage sizes that are very detrimental to the safety level of the structure, and the quality of safety can only be maintained by making such occurrence very unlikely.

SUMMARY

Exceeding the limit load has a lasting effect on safety when it causes damage. Should the damage result in loss of ultimate integrity, it becomes a threat to fail-safety if not detected. If larger damage is inflicted, the situation becomes very similar to that for damage caused by violation of any of the other Elements of Safety rules.

Unsafe environments give rise to several threats. Some must be dealt with on a 'case-by-case' basis and can result in added loads such as in 'hybrid structures', with excessively elevated temperature.

The results from failure to avoid hailstones in flight can be dealt with as 'discrete source' damage and designed for 'get-home-loads'. Undetected, 'ground-inflicted' hail damage is serious and must be dealt with in terms of inspection discipline so that the events remain rare.

The effects of excessive environments in hybrid structure come in several 'flavors':

- temporarily compromised properties;
- inflicted damage due to added 'loads';
- residual loads threatening other load cases with regular environments.

Unreported ground damage is a very serious problem which, for composites, can come with many different levels of internal damage extent and severity. Reporting before an unsafe situation has arisen is the preferred way to guard against low safety levels in service.

Example 1.4 Exceeding limit loads without any damage of permanent nature being inflicted is a temporary threat, but applies to all structures critical for the load case in question. A numerical example using practical orders of magnitude is now investigated using the assumption that 60 PSEs are involved and that each has an average of 50 detail design points (DDPs). We use the situation of a 10% exceedance of the internal limit loads involved. The probability of the joint event of a loss of limit load integrity and a simultaneous violation of the limit load condition results in

$$P(\overline{UOT_{O1}}) = 60 \cdot 50 \cdot P(\overline{B}_{EL} \mid \overline{X}D_5\overline{O}T_1) \cdot P(D_5 \mid \overline{XOT_{O1}}) \cdot P(T_{O1} \mid \overline{O}) \cdot P(\overline{O})$$

which for this example becomes

$$P(\overline{UOT_{O1}}) = 3000 \cdot 0.38 \cdot 10^{-3} \cdot 10^{-2} \cdot 0.33 \cdot 10^{-4} = 0.38 \cdot 10^{-6}$$

If we return to the discussion in Backman (2005) concerning the setting of safety level requirements, it may be remembered that:

- the basic number of unsafe flights should not exceed 1 in 100 000;
- the total share of structure 'assigned' is 10%;
- with five 'elements of safety' with equal share we then have a level of 'un-safety' that, not to be exceeded, is:

$$0.2 \cdot 10^{-6}$$

The trial value, 10^{-4}, is not quite low enough to satisfy that requirement.

It is also interesting to compare the probability of an unsafe state, which we have just arrived at, with the inherent consequence of the present-day practice of sometimes using means, μ, as the allowable value for panel strength.

A normal distribution, Φ, is assumed for the strength of composite panels, and the limit strength (remembering that 'limit' is defined as the largest load expected in service). Then the probability of being below limit strength is

$$\bar{p}_L = \Phi\left(\frac{0.67\mu - \mu}{C_V \mu}\right) = \Phi\left(\frac{-0.33}{C_V}\right)$$

Here C_V is the coefficient of variation, which for some practical values results in

C_V	T	$\Phi(t)$
0.10	-3.30	$5 \cdot 10^{-4}$
0.08	-4.12	$2 \cdot 10^{-5}$

This practice can, clearly, lead to a very poor baseline for safety. The purpose of this example is to illustrate how important even moderate exceedances of limit loads can be for safety, and how this relates to some practices presently in existence.

Another violation of operating procedures that also has grave safety consequences is the 'abuse' of not always reporting 'ground-inflicted damage'. So the probability of not reporting ground damage that violates limit integrity is the focus

$$P(\bar{U}_L \bar{O} T_{O3})$$

which can be expanded to

$$P(\bar{B}_L \mid D_i \bar{O} T_{O3}) \cdot P(D_i \mid \bar{O} T_{O3}) \cdot P(T_{O3} \mid \bar{O}) \cdot P(\bar{O})$$

Here D_i represents damage-size regions. The resulting probability is now calculated for three different regions:

i	Values in above order	Probability of occurrence
5	$10^{-1} \cdot 10^{-4} \cdot 0.33 \cdot 10^{-5}$	$0.33 \cdot 10^{-10}$
4	$10^{-3} \cdot 10^{-3} \cdot 0.33 \cdot 10^{-5}$	$0.33 \cdot 10^{-11}$
6	$1 \cdot 10^{-6} \cdot 0.33 \cdot 10^{-5}$	$0.33 \cdot 10^{-11}$

Here the numbers represent the 'probability of not reporting "ground damage" in D_5 (damage sizes for the limit requirements), D_4 (damage sizes between the requirement for ultimate and limit) and D_6 (damage sizes of extreme nature)'.

This part of the example illustrates the importance of monitoring and collecting data of unreported ground damage, especially for location of internal damage that is not accessible to 'walk-around' inspections.

1.2.4 Safety and regulations

Composites practices have evolved, for transport category aircraft, with hardly any concurrent regulations development. Instead it has often been argued that the FAR 25 and JAR 25, which primarily are specifically for metals, are also applicable to composite structure (and hybrid structure). The present state has resulted is much uncertainty, inconsistent or insufficient practices and no consistent, developing service experience.

However, as the composite structures technology has evolved further, it has become evident that special composite regulations are necessary, mainly because of all the new threats to safety, the nature of the composite materials, the development of dubious practices and the extensive need for feedback in service from both risk management and uncertainty-reduction processes.

A summary of the other elements of safety reveals that violations of procedures are detrimental when:

- properties are degraded in a way that affects design data;
- flight and ground load placards are violated and limit loads are exceeded;
- 'secondary' loads are superimposed in service or manufacture;
- damage is inflicted in a way that reduces allowable values of residual strength, or fail-safety is compromised by loss of ultimate integrity in the adjacent remaining structure.

The first role of regulation, therefore, should be to legislate and enforce the ways in which design data are produced and protected.

1.2.4.1 Safety and design values

The design data that require most attention are:

- ultimate design data
- stiffness
- fail-safe detail data
- Residual strength
- Damage resistance limits
- Maximum damage growth rates
- discrete source point design data.

Ultimate design data have been approached in a number of ways, the current practices requiring regulatory attention. There are several primary design considerations for

PSEs, i.e. the structure critical in compression, the structure critical for predominant tension loads and shear critical structures. The ultimate compression design practices often deal with at least four types of critical modes:

- open-hole compression
- compression after impact
- compression buckling
- crippling.

The ultimate tension design modes often employed are:

- filled-hole tension
- tension after impact
- residual strength.

Shear critical, ultimate design presents a much less consistent picture, but under consideration one would expect:

- open- or filled-hole shear
- shear after impact
- shear buckling and post-buckling with damage.

The 'biaxial and shear' situation, which is often the rule in today's world, would frequently require an 'equivalence, compliance demonstration based on uni-axial, internal loads data' that could be achieved by using one of the following alternatives as the criterion:

- point-design data;
- a scale-up validated algorithm using smaller test specimens;
- an empirical interaction, buckling criterion.

The effects of permanent fasteners and bolted repairs should be dealt with in terms of failure statistics for fastener installation.

The typical practice in the composites, design value world employs data of the form

$$P(\bar{B}_U S_D S_E S_S) = P(\bar{B}_U \mid S_D S_E S_S) \cdot P(S_D) \cdot P(S_E) \cdot P(S_S) \qquad (1.12)$$

Here the following events are included:

$\bar{B}_U$: design value, $D_V \leq ULR$, ultimate load requirement
S_D : state of Damage
S_E : state of environment
S_S : state of disturbance of local stress due to installation

Equation (1.12) would, for 'metal world' practices, almost exclusively deal with typical events in describing the above three states. However, in the 'composite world', practices lean toward the extreme, so, for example 'open-hole compression' is based on maximum temperature, equilibrium moisture content and an interpretation that structure adjacent to fastener holes responds as if it were an 'open hole', whether there is a permanent fastener at the location or not, accommodating potential 'bolted repairs' in some possible future time.

This formulation results in very low probabilities for the joint event described in Equation (1.12), for example for 'open-hole' compression, as we just discussed. A few situations, such as buckling without damage and with typical environmental effects, on the other hand, would produce results very similar to those with which the 'metal world' practices routinely deal.

A fair comparison of criticalities would require distinct, detailed definitions of environmental effects to consider in design, as would the effects of mechanical fasteners with different installation specifications.

The level of damage to consider in ultimate design is presently based on a prescribed energy level for a spherical impactor. In the practical world one would either have to demonstrate that that criterion covers the effects of irregular impactors at different types of trajectories with energy levels based on probability of detection, or find a type of impactor that represents 'the real world'.

The entire concept of 'regulations and criteria' for design values needs close scrutiny based on what the intended structural safety level should be. The whole concept of preserving limit integrity could be used to evaluate the criticality of the modes

$$P(\overline{U}_L) = \sum P(\overline{B}_{Li} S_{Di} S_{Ei} S_{Si})$$

(1.13)

Here the index i represents different design values per Equation (1.12).

Stiffness is an important property from an aero-elastic standpoint, from a dynamic stability standpoint and from a buckling standpoint. A very close watch has to be kept on the available tolerances of the required values. Stiffness, like strength values, is the source of many interactions between the elements of safety.

The *residual strength* of all PSEs is the baseline of limit integrity. It is an important part of structural safety, and is closely tied to the damage types and sizes used to determine limit integrity. Backman (2005) shows how the probability of an unsafe state can be expressed for a damage type:

$$P(\overline{H}_\tau U_T \overline{H}_T) = P(\overline{H}_\tau \mid D_4) \cdot P(\overline{H}_T \mid D_5) \cdot P(\overline{B}_{LT} \mid \overline{X}D_5) \cdot P(D_5 \mid \overline{X}) \cdot P(\overline{X})$$

Here we find that the second and the fourth factors on the right-hand side depend on the damage size either directly or indirectly. Thus the design requirements for damage, (regulation or approved criteria) can be derived limited to what is described in Example 1.5.

Example 1.5 We assume that orders of magnitude are tied to an airplane requirement of one unsafe flight in 100 000 flights:

$$10^{-11} = 10^{-2} \cdot x_1 \cdot 10^{-1} \cdot x_2 \cdot 10^{-2} = 10^{-5} \cdot x_1 \cdot x_2$$
$$\therefore x_1 \cdot x_2 = 10^{-6}$$

which could result in $P(\overline{H}_T \mid D_5) = 10^{-3}$ and $P(D_5 \mid \overline{X}) = 10^{-3}$.

There is enough freedom here to prescribe some verification and some criteria that could lead to a rational approach for setting requirements for damage sizes, and the type in question could imply the relation between internal and external damage sizes.

Damage resistance limits are a very important part of safety. Backman (2005) shows how an undetected 'loss of limit integrity' can lead to serious structural failures even in the size range for damage tolerance design. The objective of damage resistance must be to prevent realistic, grave threats from producing initial damage of a size appearing in the upper part of D_4 (see Figure 1.1); when maximum growth rates place the resulting damage in D_5 after only a limited number of flights, the initial damage size must be curtailed even further.

The design criteria for damage resistance therefore must depend on maximum growth rates, the extent to which the fail-safe design is validated and whether 'walk-around' inspections are possible.

Maximum damage growth rates are important and must be investigated and validated for both safe and anomalous conditions. A maximum increase of damage size during, for example, three inspection intervals, must be prescribed and validated as part of the design, safety basis. Backman (2005) contains a probability distribution that can be adapted to the a priori knowledge basis or to the updated databases.

This review of design data shows the importance of the thorough control of the requirements of the a priori knowledge base and steps to be taken to produce an ever-improved safety level in service.

1.2.4.2 Safety and design loads

Present composites practices include three different definitions of limit load.

The one appearing in FAR 25 reads:

> Limit Load is the largest load expected in service.

That often used in the damage tolerance context is:

> The maximum load experienced by the airplane in question once in a lifetime.

Another definition often used in 'Fleet Management' is:

> The maximum load experienced by the fleet in its lifetime.

It is clear that from a design standpoint an unambiguous definition of limit load is required to make the service feedback a useful contribution to safety.

The design of structure deals with internal loads and every detail design point (DDP) 'comes' with the largest internal, compression, tension and shear load during each flight. It would make sense to extract the maximum value using these three definitions, in the proper context, to determine the greatest possible values during the lifetime of the airplane. Backman (2005) gives a detailed discussion of this challenge.

Use of a definition of limit load that has a rational design influence and a consistent interpretation through the design, building and service processes of transport category airplanes is paramount for safe innovation.

1.2.4.3 Safety and residual strength

The typical situation in the 'metal world' is that *residual strength* is mostly used to determine inspection intervals. Instead fail-safety is the criterion of choice for a typical PSE, especially as the use of B-values require fail-safe detail design. Regulations and design criteria used in the 'metal world' therefore, do not have the rigor composite structures require.

The nature of composite structure is such that both *residual strength* and limit integrity are often critical considerations. In addition, fail-safety for composites is a 'major player' in compression-critical structure, causing interaction with damage resistance to become much more important. The practices employed in composite, structural design in general cannot adopt that as used with metal involving mean strength utilized as the design value for the damaged structure.

If we return to Example 1.5 we find that even if we impose very stringent probability value requirements for damage occurrence and detection, we must use B-value residual strength to satisfy the safety level indicated.

Finally the interaction between *residual strength,* damage resistance limits and maximum damage growth rates makes the formulation of residual strength criteria a complicated process with significant ramifications for structural safety.

1.2.4.4 Safety and damage resistance

Damage resistance, in the 'metal world', is a significant design activity in, for example, leading edges, nose bulkheads and windshields. However, it is not a major activity in the demonstration of fail-safe integrity. The association with fatigue cracking, crack growth and slow internal load redistribution has relegated it to a lesser importance, except for situations such as crack arrest at splice stringers, for example for large cracks in wing skins.

In the 'composites world', however, due to the fact that compression induced failures cannot be considered to involve gentle load redistributions, rather the opposite, a very 'dynamic situation, is at hand. Especially for the situation when 'walk-around' inspection is not available, large damage can result in failures at high load as limit. Dynamic damage expansion requires a *damage resistance* level that allows for failure, load redistribution and restored limit load capability in the remaining undamaged part of the structure – a total fail-safe capability that must be validated and demonstrated.

The design process for *damage resistance* has predominantly involved point design testing, which for composites involves a thorough investigation of the primary sources of the resistance and a convincing compliance demonstration process.

1.2.4.5 Maximum damage growth rates

The control of initial damage is an important part of structural design. The 'control' of *maximum damage growth rates* is also very important from both a safety and an efficiency standpoint. Small, inconsequential scales of damage can grow into severe safety threats under specific circumstances, which can involve local environments, common contaminants, local strain fields and unusual 'ground–air–ground' cycles.

The process of limiting *growth rates* requires the identification and demonstration of the significant variables. The control objective interacts with the definition of inspection intervals and an initial probability distribution for realistic growth rates must be part of both the initial design and the monitoring process. Backman (2005) shows a demonstration of growth control based on detection within three major inspection periods.

It would be expected that regulations controlling how to limit growth rates should at least identify *maximum growth rates* under the realistically 'worst' service conditions and the compliance requirements should prescribe a process for how probability of growth to 'critical size' should be developed. A rational approach to *damage growth* in composites requires both an a priori set of requirements and standards for in-service monitoring and updating, while recognizing that under certain circumstances the extent of damage does not increase and, under others, substantial growth is encountered.

1.2.4.6 Fail-safety

Structural *fail-safety* is a challenge for detail design, and totally dependent on ultimate integrity. *Fail-safety* is not a member of the list given in section 1.2.4, describing the design properties that should be given primary attention. However, it is equally important, but in an indirect way. *Fail-safety* is necessary to validate the use of design B-values. It is also a natural 'back-up' for missing 'walk-around' inspection capabilities, and a natural defense against loss of structural integrity, including 'get-home' integrity.

Fail-safety depends on the ultimate integrity of the 'remaining structure'. Thus loss of ultimate integrity does not only affect the value of the safety factor, but it degrades the *fail-safe* aspect of damage tolerance. In the 'metal world' the threats to ultimate integrity are essentially fatigue damage and crack growth, and a very common objective for the probability and confidence level of exceeding predicted life is 0.95/0.95.

In the 'composites world', we have an even more complex threat, as the 'accumulation of accidental damage' adds to the other two threats mentioned above, and load redistribution after a load path failure quite often involves violent local failure, and becomes another design consideration. The next example illustrates the orders of magnitude in the two different 'worlds'.

Example 1.6   Suppose we focus on a PSE, 'I' with $N = 50$ detail design points (DDPs), and define a safe situation as 'all the pertinent load paths having ultimate integrity preserved', then:

$$P(S_I) = P(S_{I1} \cdot S_{I2} \cdots S_{IN}) = P(S_{I1}) \cdots P(S_{IN})$$

The 'metal world' produces the following probability of an 'unsafe flight due to loss of fail-safe integrity'. It is based on the practice indicated above.

$$P(\overline{S}_I) = 1 - 0.95^{50} = 1 - 0.0769 = 0.9231$$

This obviously raises the question of how to define a safe PSE. A PSE is safe when all the DDPs are safe (by definition). A PSE fails when the structure at one DDP fails, if not designed to be fail-safe (assuming a realistic design model).

In the 'composites world', we can write the probability of failure with lost fail-safe integrity as

$$\sum_{i=1}^{2} P(\overline{AS}_{FSji}) = P(\overline{AS}_{FSj}) \tag{1.14}$$

Here a typical term can be written as:

$$P(\overline{AS}_{FSji}) = P(\overline{LYU}_{nai} R_D) + P(\overline{YU}_{nai} \overline{R}_D) \tag{1.15}$$

The first term on the right-hand side represents a 'successful redistribution of load', while the second is typical for redistribution failures. The following events are involved:

$\overline{A}$: load, $L \geq S$, the strength
$\overline{S}_{FSji}$: fail-safe integrity lost at PSE I at j with one adjacent damaged load-path
$\overline{L}$: load in the limit range
$\overline{Y}$: load-path failed
R_D : load redistribution successful

If we now suppose that the failure of the load-path takes place just after a major inspection with a period of 3000 flights, N_F, we need to expand the view to cover the whole PSE:

$$P(\overline{AS}_{FS}) = N \cdot \{2N_F \cdot P(\overline{L} \mid \overline{YU}_a R_D) \cdot P(R_D \mid \overline{YU}_a) \cdot P(\overline{Y}) \cdot P(\overline{U}_a)$$
$$+ 2P(\overline{R}_D \mid \overline{YU}_a) \cdot P(\overline{Y}) \cdot P(\overline{U}_a)\} \tag{1.16}$$

where $\overline{U}_a$ represents loss of ultimate integrity in adjacent structure.

The numerical evaluation for 3000 flights and 50 DDPs yields

$$P(\overline{AS}_{FS}) = 2 \cdot 50 \cdot \{3000 \cdot 0.333 \cdot 10^{-4} \cdot 0.9 \cdot 10^{-3} \cdot 10^{-3} \cdot 10^{-1}$$
$$+ 10^{-4} \cdot 10^{-3} \cdot 10^{-1}\}$$
$$= 50 \cdot 1.8 \cdot 10^{-8} + 10^{-6} = 1.9 \cdot 10^{-6}$$

The result emphasizes the importance of having a reliable first estimate for, and updated probabilities of, load-path failures. It also illustrates the increase in complexity in relation to composites, which can be resolved if planned and accounted for.

1.3 EFFECTS ON SAFETY AND THE INFLUENCE OF RESIDUAL STRENGTH

In the 'metallic, subsonic, transport world' it has generally been accepted that preserving structural safety is synonymous with maintaining damage tolerance integrity.

In most cases this is interpreted as protecting the residual strength limits required from fatigue, corrosion and accidental damage.

Entering the 'composites world' has widened our threat focus to encompass the following concerns:

- ultimate integrity and fail-safety;
- damage tolerance integrity;
- damage resistance limits;
- damage growth limits;
- stiffness preservation.

The study of elements of safety in previous sections has revealed the following interactions both before 'roll-out' and in service:

Table 1.1 Interaction between elements of safety

Cause Effect	Elements of safety influences				
	Design	Manufacture	Maintenance	Operation	Requirements
Property reduction	•	•	•	•	•
Damage	•	•	•	•	–
Secondary loads	•	•	•	–	–
Load exceedance	•	–	–	•	•
Inspection failure	–	•	•	•	–
Damage reporting failure	–	–	•	•	–

Here a dot (•) indicates 'participation' and a bar (–) 'not involved'. It is clear that the elements of safety require a high level of interaction, and a balancing of burden is clearly a large part of safety management.

1.3.1 Safety and integrity

Safety depends on the preservation of structural integrity. Equation (1.17) illustrates the relation:

$$P(\overline{AU}) = P(\overline{A} \mid \overline{U}) \cdot P(\overline{U}) \le P(\overline{U}) \tag{1.17}$$

Backman (2005) shows several, practical illustrations of situations when the above equation and inequality obeys

$$P(\overline{A} \mid \overline{U}) \rightarrow \text{ when } n \rightarrow N \text{ in the range } N \ge 3000$$

The threats to integrity arise with damage scenarios. We will start with a damage scenario that does not include growth (maximum growth rate is demonstrated as ~0),

1.3.1.1 'No-growth' scenario (Figure 1.6)

This scenario can be expressed as:

$$P(\overline{X}_\tau D_{4\tau} \overline{H}_\tau S_{u\tau T} D_{4T} \overline{U}_T \overline{H}_T) = P(\overline{H}_T \mid D_{4T}) \cdot P(\overline{U}_T \mid D_{4T}) \cdot P(S_{u\tau T})$$
$$\cdot P(\overline{H}_\tau \mid D_{4\tau}) \cdot P(D_{4\tau} \mid \overline{X}_\tau)$$
$$\cdot P(D_{4T} \mid \overline{X}_\tau) \cdot P(\overline{X}_\tau) \tag{1.18}$$

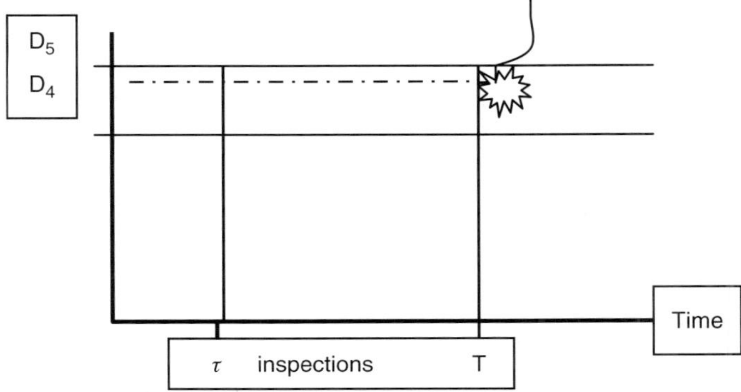

Figure 1.6 Damage with no growth, undetected.

A realistic, practical assessment of this scenario (no growth–loss of integrity–no detection), might result in the following probability value for an unsafe state $P(\bar{S}_T)$:

$$P(\bar{S}_T) = 10^{-2} \cdot 10^{-2} \cdot 0.5 \cdot 10^{-2} \cdot 10^{-2} \cdot 10^{-2} = 0.5 \cdot 10^{-10}$$

It is interesting to see what 'walk-around' inspections would produce even for a low probability of detection of 0.5:

No. flights N	Prob. of not being detected in n flights	Prob. of being detected before or after n flights
10	10^{-3}	0.999
20	~0	~1

Making accommodation for 'walk-around' inspections must clearly be a high priority in design.

1.3.1.2   Growth scenario (Figure 1.7)

A safe design needs a validation that the 'maximum growth rate' is negligible. This leaves us with many situations wherein growth must be considered in producing safe design. Figure 1.7 shows a growth case that if not detected will lead to an 'unsafe state' and failure. The 'natural' way to deal with this case is to establish a damage resistance of the structure so that 'design threats' only produce damage of moderate scale and 'reasonable' growth rates, thereby making detection a realistic solution.

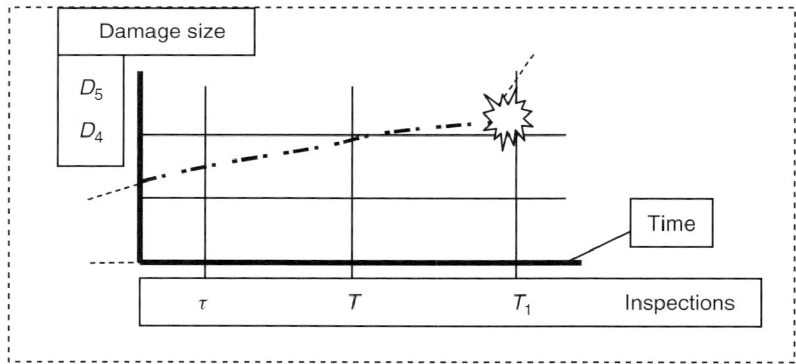

Figure 1.7 Damage growth and no walk-around inspection

The probability value of interest for this scenario is the 'probability of an unsafe state' at T, $P(\bar{S}_T)$

$$P(\bar{S}_T) = P(\bar{X}_\tau D_{4\tau} \bar{H}_\tau E_{\iota T} D_{5T} \bar{U}_T \bar{H}_T). \tag{1.19}$$

This equation can be expanded to

$$\begin{aligned} P(\bar{S}_T) = {}& P(\bar{H}_T \mid D_{5T}) \cdot P(\bar{U}_T \mid D_{5T} E_{\tau T}) \cdot P(\bar{H}_\tau \mid D_{4\tau} \bar{X}_\tau) \cdot P(D_{5T} \mid E_{\tau T}) \\ & \cdot P(E_{\tau T}) \cdot P(D_{4\tau} \mid \bar{X}_\tau) \cdot P(\bar{X}_\tau) \end{aligned} \tag{1.20}$$

Here we deal with the following events:

$$\begin{aligned} \bar{S}_T \quad & : \quad \text{unsafe state at time } T \\ \bar{H}_T \quad & : \quad \text{undetected at time } T \\ D_{5T} \quad & : \quad \text{damage size in } D_5 \text{ at } T \\ \bar{U}_T \quad & : \quad \text{loss of integrity at } T \\ E_{\tau T} \quad & : \quad \text{survival between } \tau \text{ and } T \\ \bar{H}_\tau \quad & : \quad \text{undetected at } \tau \\ D_{4\tau} \quad & : \quad \text{in size region 4 at } \tau \\ \bar{X}_\tau \quad & : \quad \text{damage present at } \tau \end{aligned}$$

A realistic assessment of the scenario in terms of practical requirements yields

$$P(\bar{S}_T) = 10^{-3} \cdot 10^{-1} \cdot 10^{-2} \cdot 0.3 \cdot 0.9 \cdot 10^{-2} \cdot 10^{-2} = 0.27 \cdot 10^{-10}$$

It is important that inspection quality, residual strength design value choices, growth rates, damage resistance limits and damage probabilities interact to achieve a desirably low probability value: a safe design.

1.3.1.3 Damage scenario with 'walk-around' inspection (Figure 1.8)

Figure 1.8 illustrates large-scale damage inflicted at time t and detected by 'walk-around' inspection after n flights. The following numerical example illustrates orders of magnitude:

Assume that the probability of detection at a random flight is 0.5.
Then the probability of detection within 30 flights is $p_{d30} = 1 - 0.5^{30} \approx 1.0$

The importance of making 'walk-around' possible is clearly an important part of detail design.

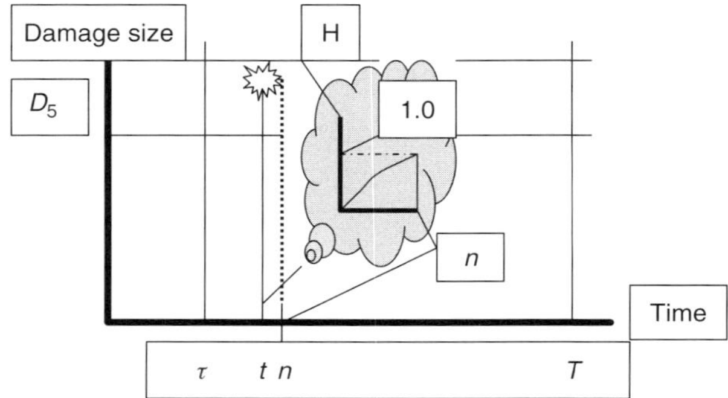

Figure 1.8 Damage soon after a major inspection and found by 'walk-around'.

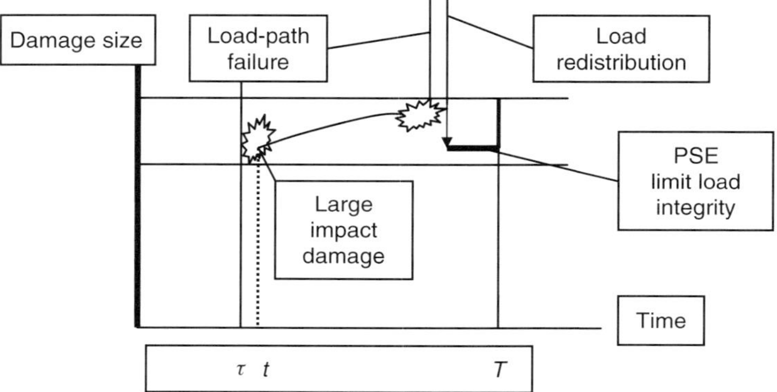

Figure 1.9 Damage scenario with large-scale damage and load-path failure.

1.3.1.4   Damage scenario with no 'walk-around' and fail-safe (Figure 1.9)

This scenario involves large damage (e.g. $d_s \ni D_4$), inflicted at t (just after a major inspection at τ), damage growth, load-path failure, load redistribution and a preserved limit load integrity. The probability of this joint event, F_E, can be expressed as

$$P(F_E) = P(\bar{X}_t D_{4t} D_{5T} \bar{Z}_{tT} R_D U_{UtT} U_{LT} H_{TZ})$$

(1.21)

Here the participating events can be described as

$\bar{X}_t$: damage present at time t

D_{4t} : damage size in region 4 at t

D_{5T} : damage size growth to region 5 by T

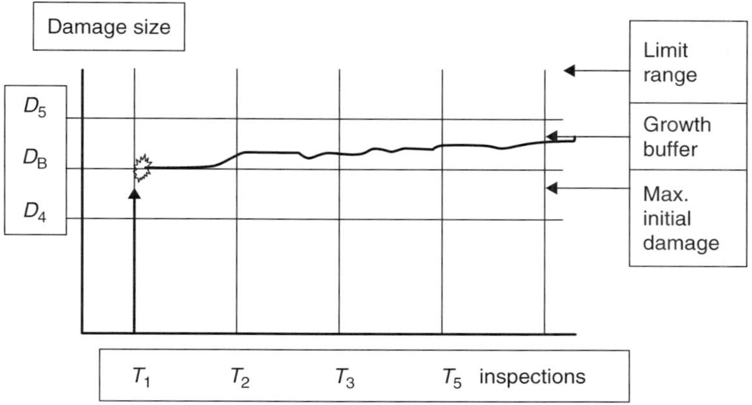

Figure 1.10 Damage resistance and growth limits.

$\overline{Z}_{tT}$: load-path fails between t and T

R_D : internal loads successfully redistributed

U_{UtT} : ultimate integrity of adjacent structure preserved

U_{LT} : limit integrity of the PSE preserved

H_{TL} : the failed load-path is discovered at T

Equation (1.21) can be expanded:

$$P(F_E) = P(H_{TZ} \mid U_{LT}U_{UtT}\overline{Z}_{tT}R_D) \cdot P(U_{LT} \mid U_{UtT}\overline{Z}_{tT}R_D) \cdot P(R_D \mid U_{tT}\overline{Z}_{tT})$$
$$\cdot P(U_{UtT}) \cdot P(\overline{Z}_{tT} \mid U_{UtT}) \cdot P(D_{5tT} \mid D_{4t}\overline{X}_t) \cdot P(D_4 \mid \overline{X}_t) \cdot P(\overline{X}_t) \quad (1.22)$$

The following is a numerical illustration of what realistic values, interpreted from the standpoint of 'practical world' requirements, become:

$$P(F_E) = 0.90 \cdot 0.90 \cdot 0.80 \cdot 0.90 \cdot 10^{-5} \cdot 0.3 \cdot 10^{-2} \cdot 10^{-2} = 10^{-10}$$

The four scenarios just described all have very important ties between integrity and damage tolerance and constitute an important set of requirements in the quest for safe composite structure.

Two closely related concepts, damage resistance and damage growth rate, are both 'drivers of safe designs' and are explored in the next section.

1.3.1.5 Scenario – damage resistance and damage growth (Figure 1.10)

Figure 1.10 illustrates the advantage of having a 'designed-in' buffer zone in the damage size regions to allow for maximum growth in three inspection intervals (example of criterion), without entering the 'limit load damage interval'. It also explains the need to include a damage resistance, maximum, initial damage size interval for a defined maximum design threat. Thus in conclusion it is inferred that 'damaged-based designs in the "composite world"' require threat evaluation that includes:

- damage tolerance
- damage resistance
- maximum growth rates.

In order to produce safe structures, it is recommended that the design process includes safety-based design constraints. The other element processes involved should also have requirements for what is considered safe manufacturing, safe maintenance, safe operation and safe requirements formulations from the standpoint of structural integrity both in formulating detail designs and in identifying the unacceptable range of values for different defects. The measure for control of unacceptable defects values could be based on their contribution to an unsafe state:

$$P(\bar{X}_M T_i V_{ij}) = P(V_{ij} \mid T_i \bar{X}_M) \cdot P(T_i \mid \bar{X}_M) \cdot P(\bar{X}_M)$$

Here the first factor could be used to control the range if

$\bar{X}_M$: defective manufacturing process
T_i : type of defect
V_{ij} : value range for defect type i

This definition of the probability of an integrity-threatening defect will be used throughout the book. In the next chapter we will study the effects of defects caused by different elements.

1.4 CONCLUSIONS

The contributions of all the elements of safety to the probability of an unsafe state are of the same nature and all of the elements are of equal importance to structural safety.

Each of the elements of safety must be controlled by processes that render the occurrence of realistic defects unlikely; extreme values of the defects belong out with the definition of 'safe elements' (e.g. six sigma).

All elements include scenarios that produce results which make fail-safety and 'walk-around' inspections necessary back-ups to safe states for load-paths in most principal structural elements (PSEs).

Damage tolerance and limit integrity are the dominating considerations in structural safety for composite materials, and as a consequence the following properties have a very strong interaction in structural safety:

- damage resistance;
- damage maximum, growth rates;
- damage tolerance.

The aspects of safety come in several combinations:

- Limit integrity;
- Mechanical damage;
- Damage tolerance;
- Residual strength;
- Detection;
- Safe state;
- Failure after a period of lost integrity;
- Walk-around.

Fail-safe integrity, hard-to-detect defects, ultimate, internal loads after load path failure, and damage must be detected or failure will occur.

Ultimate integrity, B-values require fail-safety.

Chapter 2
Structural Integrity and Safety Threats

Backman (2005) describes the role of structural integrity in structural design, and emphasizes the importance of structural limit integrity for composites. Structural design, to a large extent, therefore focuses on accidental damage and the use of appropriate processes, methods and approaches in order to achieve total structural safety. *Safety Management* identifies the additional 'elements of safety' and their roles. These elements are:

- manufacturing
- maintenance
- operation
- requirements formulation.

The breakdown of the manufacturing element into the parts that constitute important targets for safety was given in Chapter 1. The remaining elements can be factored in the following manner.

Maintenance can be factored as

$$S_I = S_{IS} \cdot S_{II} \cdot S_{IR}$$

Here the first factor represents safe scheduled maintenance; the second, safe inspection; and the third, safe repair.

Operation can be expanded as

$$S_O = S_{OL} \cdot S_{OE} \cdot S_{OR}$$

The first factor represents safe operational loads; the second factor, a safe environment; and the third, a safe inspection and reporting procedures.

Requirements formulation can be expressed as

$$S_R = S_{RR} \cdot S_{RC} \cdot S_{RE}$$

The first factor represents safe regulations; the second, safe design criteria; and the third, safe education.

Safety management investigates the effects on safety when mishaps give rise to results outside prescribed limits and identifies the actions required to maintain safety.

The previous chapter analyzed the nature of these deviations. The nature of the threat was found to be concentrated in three 'types'. These are:

- unacceptable structural property reductions;
- digressions beyond limit loads;
- spurious and/or unreported damage.

Table 1.1 summarizes effects in terms of cause–effect relations, the data given representing the 'end-results', independent of cause or duration. This chapter discusses how these three threats affect safety in terms that would be compatible with the monitoring of the structure in service; analyses of performance; and updating of safety levels. It also investigates the corrective actions required during service and for long-term, systemic improvements in design, process control, quality control, inspection quality, repairs and reporting, and it discusses the limits imposed by threats to structural integrity.

If we return to the concept that an 'undetected loss of integrity is unsafe' the probability of an 'unsafe state' can be written as

$$P(\bar{S}_T) = P(\bar{X}_E T_i V_i \bar{B}_T D_{5T} \bar{H}_T \bar{H}_\tau D_{4\tau})\qquad(2.1)$$

Here the events are:

$$
\begin{array}{ll}
\bar{X}_E & : \text{element of safety involved in the mishap} \\
T_i & : \text{type of mishap} \\
V_i & : \text{degree of mishap} \\
\bar{B}_T & : \text{strength is less than requirement} \\
D_{4\tau}\ \&\ D_{5T} & : \text{damage size region 4 at } \tau \text{ and 5 at } T \\
\bar{H}_\tau\ \&\ \bar{H}_T & : \text{damage undetected at } \tau \text{ and } T.
\end{array}
$$

We will now study the effects of the different elements of safety in detail.

2.1 INTEGRITY AND MANUFACTURING

The failure in composites manufacturing that often receives the greatest attention is laminate processing mishaps and the resulting changes in design values, but there is an entire family of properties that will affect safety. We will review the integrity requirements for different scenarios and their role in managing safety. An example of the definition of the probability of loss of manufacturing integrity is derived from the definition of the probability of safe manufacturing:

$$P(M \mid IOR) = P[P_R \cup (\bar{P}_R\,HC)]\qquad(2.2)$$

Here,

M : safe manufacturing
I : safe maintenance
O : safe operation
R : safe requirements
P_R : safe process
H : detection
C : correction of mistakes in process.

The corresponding probability for unsafe manufacturing is

$$P(\bar{M} \mid IOR) = P(\bar{P}_R \overline{HP}_{RF}) + P(\bar{P}_{RF}\bar{C}H\bar{P}_{RI}) \tag{2.3}$$

Here:

$\bar{P}_{RI}$: an initial process failure
$\bar{P}_{RF}$: a failed correction process.

Equation (2.3) describes the probability of unsafe manufacturing given safe remaining elements. We are now investigating the different effects of manufacturing process failures. There are a number of characteristics unique to the defects caused by manufacturing. They are material and process-specific, so different materials have to be approached in different ways. Reduction of structural properties due to process discrepancies can come in many combinations affecting different types of properties. The specifics must be part of material and process characterizations. This can affect only a very limited area, and can involve sub-regions of a principal structural element (PSE), or compromise the whole PSE. The resulting defects can affect structural integrity and constitute a permanent presence. A defect can be one that must be combined with other errors such as temporarily exceeding the limit load.

The loss of fail-safe integrity is a troublesome situation that is corrected only by avoidance or prevented by utilizing an ultimate positive margin of safety.

2.1.1 Property reductions

Equation (2.1) can be expanded in the following manner:

$$\begin{aligned} P(\bar{S}_T) = {}& P(\bar{H}_T \mid D_{5T} \bar{X}_M T_1 V_i) \cdot P(\bar{B}_T \mid D_{5T} \bar{X}_M T_1 V_i) \cdot P(D_{5T} \mid D_{4T} \bar{X}_M T_1 V_i) \\ & \cdot P(\bar{H}_T \mid D_{4T} \bar{X}_M T_1 V_i) \cdot P(D_{4T} \mid \bar{X}_M T_1 V_i) \cdot P(V_i \mid T_1 \bar{X}_M) \\ & \cdot P(T_1 \mid \bar{X}_M) \cdot P(\bar{X}_M) \end{aligned} \tag{2.4}$$

We now investigate a numerical example in the range of moderate properties.

Example 2.1 We now assume normally distributed (Φ), residual-strength variables, and a 10% property reduction of B-value design data. We also assume a coefficient of variation, $C_V = 0.10$, and we can write the probability of residual strength being less than the requirement at T, as

$$P(\bar{B}_T \mid D_{5T}\bar{X}_M T_1 V_i) = \Phi\left(\frac{0.87\mu - 0.90\mu}{C_V\,0.90\mu}\right) = \Phi(-0.3) = 0.38$$

The probability of an unsafe state then becomes

$$P(\bar{S}_T) = 10^{-2} \cdot 0.38 \cdot 10^{-2} \cdot 10^{-1} \cdot 10^{-1} \cdot 0.3 \cdot 0.3 \cdot 10^{-4} = 0.4 \cdot 10^{-11}$$

The numerical value, when compared against common requirements, shows that stringency needs in process control, quality assurance and quality control can be compelling in terms of safety. More serious degradation of properties fast becomes unmanageable. It is, however, the last factor, the probability of 'manufacturing defects' occurring, where the need for control is largest. It controls orders of magnitude of the probability of an unsafe state.

A review of the commonly expected types of process failures includes:

T_1 : laminate processing mishaps (e.g. fiber misalignment);
T_2 : co-bonding/co-curing mishaps (e.g. reduced bond strength);
T_3 : installation mishaps (e.g. mechanical fastener errors);
T_4 : assembly mishaps (e.g. miss-drilled holes);
T_5 : accidental damage (e.g. tool drops).

Of these types the first two have predominantly 'reduced property' effects. Detection is often difficult and fail-safety, including surviving with one failed load-path, must be considered. The effects are reductions of:

- strength and/or stiffness
- damage tolerance
- damage resistance
- damage growth rates
- fatigue resistance (increase in damage accumulation).

The first two items on the list involve three different integrities. The undetected loss of limit integrity was the focus of Example 2.1. The loss of limit integrity can be written as

$$P(\bar{U}_L) = P(\bar{B}_L\bar{X}) = \sum_{i=3}^{6} P(\bar{B}_L\bar{X}D_i) \tag{2.5}$$

Here the events are:

$\overline{U}_L$: loss of limit integrity
$\overline{B}_L$: strength is less than requirement
$\overline{X}$: damage is present
D_i : damage size regions $i = 3,...,6$

This covers the 'limit residual strength' definition. The fact that ultimate integrity also can cause the loss of fail-safe integrity will now be investigated. The loss of ultimate integrity alone can be a reason for this. The probability of loss of fail-safety can be written as

$$P(\overline{U}_{FT}) = P(\overline{B}_{Lit}\,\overline{B}_{uat}\,\overline{X}_{at}D_{2ai}\,\overline{X}_{it}D_{5it}\,\overline{Y}_{iT}) \tag{2.6}$$

This equation describes loss of limit strength in load path i, loss of ultimate strength in the adjacent structure and a load-path failure at T.

The participating events are:

$\overline{U}_{FT}$: loss of fail-safety at T
$\overline{B}_{Lit}$: loss of limit strength in load-path i at t
$\overline{B}_{uat}$: loss of ultimate strength in adjacent structure at t
$\overline{X}_{at}$: damage present in adjacent structure at t
D_{2at} : damage size in region D_2 for adjacent structure at t
$\overline{X}_{it}$: damage present in load-path i
D_{5it} : damage size in load path in region 5 at t
$\overline{Y}_{iT}$: load-path i fails at T.

Equation (2.6) can be expanded in the following manner:

$$P(\overline{U}_{FT}) = P(\overline{B}_{Lit}\,|\,\overline{X}_{it}D_{5it}) \cdot P(D_{5it}\,|\,\overline{X}_{it}) \cdot P(\overline{X}_{it}) \cdot P(\overline{Y}_{iT}\,|\,\overline{B}_{Lit}\,\overline{X}_{it}D_{5it})$$
$$\cdot\, P(\overline{B}_{uat}\,|\,\overline{X}_{at}D_{2at}) \cdot P(D_{2at}\,|\,\overline{X}_{at}) \cdot P(\overline{X}_{at}) \tag{2.7}$$

We will now use equation (2.7) to assess numerical values of undetected loss of fail-safe integrity. In addition there is the threat of structural property reductions to consider.

Example 2.2 We now select requirement-driven values as a baseline for the probability of loss of fail-safe integrity, yielding the following assessment:

$$P(\overline{U}_{FT}) = 10^{-1} \cdot 10^{-3} \cdot 10^{-2} \cdot 10^{-1} \cdot 10^{-1} \cdot 0.5 \cdot 2 \cdot 10^{-2} = 10^{-10}$$

We will now evaluate the 'unsafe state', undetected loss of integrity:

$$P(\overline{S}_T) = P(\overline{U}_{FT}) \cdot P(\overline{H}_{T_1} \mid \overline{Y}_{iT}) = 10^{-10} \cdot 10^{-3} = 10^{-13}$$

It should be noted that this represents the probability for a detail design point (DDP).

Thus, consequently it has been shown that this type of manufacturing mishap can involve the three different integrities, i.e. limit integrity, ultimate integrity and fail-safe integrity. The secret to neutralizing this kind of threat lies in avoidance. The implementation of process control, quality assurance and quality control must be such that these effects are associated with low probabilities.

Another less obvious property reduction is associated with reduced damage resistance. The dominating effect is that damage sizes for threats of given energy levels become larger. The definition of maximum initial damage sizes, for designing to preclude accidental damage in safe structure, was previously described as not exceeding the sizes in region D_4 for moderate, practical growth rates. It also contained a modification with a 'buffer zone' for fast growth rates.

We will focus initially on moderate growth rates. In this case we find that the troubling consequence could involve initial damage in region D_5, or in the upper range of D_4, both resulting, potentially, in a substantial number of flights with 'large damage' before the next major inspection.

The probability of an 'unsafe state' caused by this process violation and ensuing large-scale accidental damage can be written as

$$P(\overline{S}_T) = P(\overline{X}_M T_i V_i \overline{Y}_t D_{kt} D_{lT} \overline{U}_T \overline{H}_T) \tag{2.8}$$

This equation includes the joint event characterized by a 'manufacturing defect', $\overline{X}_M$, of type T_i of degree V_i, an accidental impact at t, a loss of 'limit integrity' between t & T and non-detected damage at T. Here D_{kt} represents the initial damage size and D_{lT} the size region at the inspection at T. We will use Example 2.3 to illustrate the different influences involved.

Example 2.3 Potentially the property reduction can reduce residual strength, increase damage size and reduce the probability of damage sizes. We assume no growth, in this example, so the reduction in damage resistance will be the main effect. An expansion of Equation (2.8) can take the form:

$$P(\overline{S}_T) = P(\overline{H}_T \mid \overline{U}_T D_{5t} \overline{Y}_t) \cdot P(\overline{U}_T \mid D_{5t} \overline{Y}_t) \cdot P(D_{5t} \mid \overline{Y}_t \overline{X}_M T_i V_i) \cdot P(\overline{Y}_t)$$
$$\cdot P(V_i \mid T_i \overline{X}_M) \cdot P(T_i \mid \overline{X}_M) \cdot P(\overline{X}_M) \tag{2.9}$$

The crucial factors influenced by the property reduction are the second factor and the third factor. This makes the residual strength, for a given damage size, a target for safety monitoring, and so is the initial, damage size due to impact in the severe range.

The numerical values likely at this situation yield

$$P(\overline{S}_T) = 10^{-3} \cdot 0.4 \cdot 10^{-2} \cdot 10^{-2} \cdot 0.3 \cdot 0.4 \cdot 10^{-4} = 0.5 \cdot 10^{-12}$$

It is noteworthy that the critical value for reaching the above probability level is the probability of manufacturing process failures which here has been driven down to 10^{-4}.

The notable changes from 0.10, for uncompromised processing, to 0.4 for this type of mishap, and the healthy value for the damage size probability of 10^{-3} to the compromised 10^{-2} do not by themselves present any significant challenge, provided a vigilant control and assurance program is in place.

The remaining effect, increased growth rate, can, when influenced by property reduction, produce a similar result to that seen in Example 2.3. The special case of increased damage growth rate can be expressed as

$$P(\overline{S}_T) = P(\overline{H}_T \,|\, D_{5T}\overline{U}_T) \cdot P(\overline{U}_T \,|\, D_{5T}) \cdot P(D_{5T} \,|\, D_{4t}\overline{X}_M\overline{Y}_tT_iV_i)$$
$$\cdot P(D_{4t} \,|\, \overline{Y}_t\overline{X}_MT_iV_i) \cdot P(\overline{Y}_t) \cdot P(V_i \,|\, T_i\overline{X}_M) \cdot P(T_i \,|\, \overline{X}_M) \cdot P(\overline{X}_M) \qquad (2.10)$$

The numerical evaluation, in comparison with Equation (2.9), yields the following results:

$$P(\overline{S}_T) = 10^{-3} \cdot 10^{-1} \cdot 0.9 \cdot 0.2 \cdot 10^{-2} \cdot 10^{-2} \cdot 0.3 \cdot 10^{-4} = 0.5 \cdot 10^{-13}$$

This result, not surprisingly makes Equation (2.9) the critical consideration.

2.1.1.1 Damage accumulation

'Composites' represent a family of structural materials for which the accumulation of structural damage is a complicated process. It involves fatigue damage initiation, damage growth contributions, continued accidental damage occurrences and, in addition, the contributions from property changes due to materials and manufacturing process failures.

The effects created by manufacturing mishaps, and considered in this context, are manufacturing flaws (mechanical damage), reduced damage resistance, increased damage growth rates and reduced residual strength for given damage sizes. The field of multi-site damage and damage design criteria require a foundation different from the accepted practice of using load enhancement factors, in the 'metal world'.

The dominating concerns in the context of 'multi-site damage' are cascading interaction of damage sites or the progressive loss of 'fail-safe integrity'. The fail-safe integrity of the PSE is lost when one or more of the 'adjacent' load-paths lose their ultimate integrity as internal loads cannot be redistributed from a failed load-path. If we now investigate zero margin of safety structure we have the following situations at T, brought about by a combination of damage sources and property reductions:

$$\text{Manufacturing flaws: } U_{UMP} = B_{UT}\,\overline{X}_{MU}\,\overline{D}_{UT}$$

$$\text{Accidental damage: } U_{UMA} = B_{UT}\,\overline{X}_{MU}\,\overline{X}_{Ui}\,\overline{D}_{UT}$$

$$\text{Fatigue damage: } U_{UMF} = B_{UT}\,\overline{X}_{MU}\,\overline{X}_{UFt}\,\overline{D}_{UT}$$

$$\text{Property degradation: } U_{UMD} = B_{UT}\,\overline{X}_{MU}\,X$$

These definitions of preserved ultimate integrity at T deal with two events. The situation when manufacturing mishaps have taken place and when damage has been inflicted and subsequently grows. The quantities that are the targets for our safety management are: the probability of loss of ultimate integrity due to manufacturing flaws; accidental damage and fatigue damage; and property degradation due to process failures. A useful approximation can be written as

$$P(\overline{U}_{UT}) = 1 - P(U_{UMP} \cdot U_{UMA} \cdot U_{UMF} \cdot U_{UMD})$$
$$\approx [P(\overline{U}_{UMF}) + P(\overline{U}_{UMA}) + P(\overline{U}_{UMF}) + P(\overline{U}_{UMD})] \qquad (2.11)$$

The probability of loss of ultimate integrity at T due to manufacturing mishaps producing property reductions is

$$P(\overline{U}_{UMP}\,\overline{X}_{MU}) = P[\overline{B}_{UT}\,\overline{X}_{MU}\,\overline{D}_{UT}(T_{V1} \cup T_{V2})]$$

Here $T_{Vi} = T_i\,V_i$ represents 'type of effect' and 'value', and the equation describes the joint event; the probability of loss of ultimate integrity with manufacturing process violation caused 'property reductions'.

The probability of loss of ultimate integrity at T due to accidental damage at i is

$$P(\overline{U}_{UMA}\,\overline{X}_{MU}) = P[\overline{B}_{UT}\,\overline{X}_{MU}\,\overline{X}_{Ui}\,\overline{D}_{UT}(T_{V1} \cup T_{V2})]$$

The probability of loss of ultimate integrity at T due to fatigue damage at i is

$$P(\overline{U}_{UMF}\,\overline{X}_{MU}) = P[\overline{B}_{UT}\,\overline{X}_{MU}\,\overline{F}_{Ui}\,\overline{D}_{UT}(T_{V1} \cup T_{V2})]$$

The probability of loss of ultimate integrity at T due to property degradation without mechanical damage is

$$P(\overline{U}_{UMD}\,\overline{X}_{MU}) = P[\overline{B}_{UT}\,\overline{X}_{MU}\,X(T_{V1} \cup T_{V2})]$$

This leads to the typical form of loss of integrity when manufacturing process violations are considered and when additional events during service worsen the situation. The equation below in principle describes the probabilities of inadequate residual strength level due to manufacturing flaws, accidental damage in service, fatigue damage or property degradation causing damage size at time T that is larger than the maximum, ultimate design requirement. The expression for types and magnitudes is replaced by T_V:

$$P(\bar{U}_{UMF}\,\bar{X}_{MU}\,T_V) = P(\bar{B}_{UT} \mid \bar{X}_{MU}\,\bar{F}_{Ut}\,\bar{D}_{UT}\,T_V) \cdot P(\bar{D}_{UT} \mid \bar{X}_{MU}\,T_V\,\bar{F}_{Ut})$$
$$\cdot\, P(\bar{F}_{UT} \mid \bar{X}_{MU}\,T_V) \cdot P(T_V \mid \bar{X}_{MU}) \cdot P(\bar{X}_{MU})$$

The following scenarios, involving a basic manufacturing process failure, all lead to loss of ultimate integrity and, as a consequence, a loss of fail-safe integrity:

- property reduction;
- property reduction and mechanical flaws;
- property reduction and accidental damage in service;
- property reduction and fatigue damage initiation.

We now return to the formulation of chapter 1, Equation (1.2), where in the second factor deals with 'manufacturing' and is

$P(S_M \mid S_I S_O S_R)$ resulting in a contribution to an unsafe state that is:

$P(\bar{S}_M \mid S_I S_O S_R)$ limiting the contributions to what has been discussed above　　(2.12)

Thus the damage resistance reductions, growth rate increases, fatigue resistance losses and property reductions due to manufacturing 'mishaps' will be the primary effects in the accumulation of damage.

Example 2.4　Suppose Type 1 and Type 2 manufacturing mishaps are the focus, and the values of the reductions are 10%. The purpose of this example is to illustrate the significant contributions to the probability of loss of ultimate integrity. We assume both sets of data are normally distributed and $C_V = 0.10$, and that the pristine set is represented by B-values. Figure 2.1 illustrates a way to consider the effect of property reduction.

The amount of reduction is k

k	$\Phi\!\left(\dfrac{s-\mu}{C_V\mu}\right)$	s	t	$\Phi(t)$
1.0	0.10	0.87μ	-1.3	0.100
0.9	–	0.87μ	-0.3	0.380
0.8	–	0.87μ	0.9	0.820
0.7	–	0.87μ	2.4	0.992

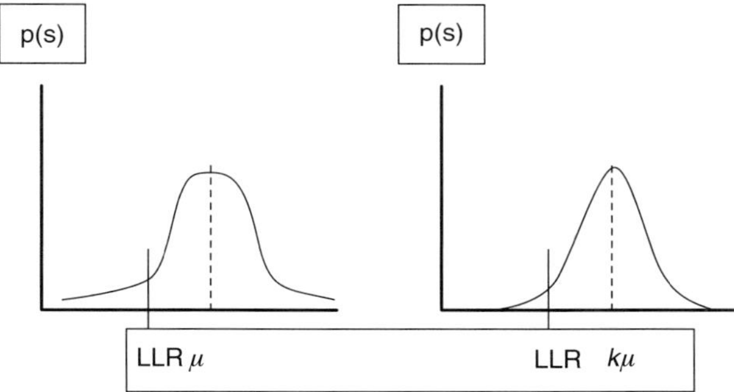

Figure 2.1 Pristine and reduced properties – a comparison.

It is clear from the resulting numerical probability values that this amount of property reduction can be troublesome. We will now evaluate the following equation:

$$P(\bar{U}_{UMP}\bar{X}_{MU}) = P[\bar{B}_{UT}\bar{X}_{MU}(T_{V1} \cup T_{V2})]$$

We assume property reductions without any mechanical damage, and the special case for Type 1 becomes:

$$P(\bar{U}_{UMP}\bar{X}_{MU}) = P(\bar{B}_{UT} \mid \bar{X}_{MU}T_1V_1)P(V_1 \mid T_1\bar{X}_{MU})P(T_{1\mid}\bar{X}_{MU})P(\bar{X}_{MU})$$
$$= 0.38 \cdot 10^{-2} \cdot 0.2 \cdot 10^{-6} \approx 10^{-9}$$

We now return to looking at how present-day safety objectives would influence the 'design requirements'. Under President Clinton, Vice-president Gore's Commission on Safety of Aviation made it clear that existing vehicle performance establishes the standards as:

> One unsafe flight in 100 000 resulting in a probability for design: 10^{-5}
> The improvement objective for ~2015 is one order of magnitude: 10^{-6}
> The part assigned to structure is 10%: 10^{-7}
> Each element of safety is assigned 20%: $2 \cdot 10^{-8}$
> We now assume 50 PSEs: $4 \cdot 10^{-10}$

The type of process failure that can be expected for Types 1 and 2 can often be expressed relative to the probability of a safe state of a PSE with 50 detail design points (DDPs) as:

$$P(S_P) = P(S_{D1}S_{D2}...S_{Dn}) \Rightarrow \text{in general } P(\bar{S}_P) = \sum_{1}^{n} P(\bar{S}_{Di})$$

Here, if many DDPs have defects, then failure takes place and detection in process control or quality control must occur.

These equations illustrate the importance of a detailed understanding of the manufacturing process deviation. The special case with a 'global' PSE process failure results in common process defects for all DDPs:

$$\text{We assume 50 PSEs: } 4 \cdot 10^{-10}$$

For the general case:

$$\text{We assume 50 DDPs: } 8 \cdot 10^{-12}$$

The final values clearly depend on the number of units for the fleet, and it is clear that the value of the factor $P(\overline{X}_{MU}T_V)$ can be used to bring the probabilities into compliance.

It is of great importance for requirement definitions that the nature of process failures can be identified as involving a total PSE (e.g. process failures in autoclaves involving co-bonding), or only affecting local quality (e.g. local resin 'starvation').

The target setting also involves the maximum value of the resulting effect (e.g. 20% property reduction). One important threat involves the reduction of **ultimate strength** and, as a consequence, loss of fail-safe integrity, which can defeat the purpose of fail-safety for the total service life if not corrected. For permanently lost integrity, limit load becomes a serious threat. The following equation illustrates the point:

$$P(\overline{A}\,\overline{U}_L\overline{X}_M T_i V_i) = P(\overline{A}\,|\,\overline{U}_L\overline{X}_M T_i V_i) \cdot P(\overline{U}_L\,|\,\overline{X}_M T_i V_i) \cdot P(V_i\,|\,\overline{X}_M T_i)$$
$$\cdot\, P(T_i\,|\,\overline{X}_M) \cdot P(\overline{X}_M) \tag{2.13}$$

$$
\begin{array}{lll}
\overline{A} & : & \text{strength, } s \leqslant \text{load, } l, \text{ failure} \\
\overline{U}_L & : & \text{limit integrity lost} \\
\overline{X}_M & : & \text{manufacturing-caused property reduction} \\
T_i & : & \text{type of manufacturing mishap} \\
V_i & : & \text{amount of reduction}
\end{array}
$$

We now assume a 30% reduction in strength due to a manufacturing mishap. We also assume a normal distribution with $C_V = 0.10$

$$\Phi\left(\frac{\dfrac{0.87\mu_x}{1.5} - \mu_x}{0.10 \cdot \mu_x}\right) = \Phi(-4.2) = 10^{-5}$$

Here the B-value gives rise to $0.87 \cdot \mu_x$, and the reduction results in

$$\Phi\left(\frac{0.87\mu_x - 0.7\mu_x}{0.10\mu_x}\right) = 0.99$$

Equation (2.13) is now used to establish the requirement; the probability of failure is 1 because this type of reduction is rarely detectable and the definition of limit load is such that it will occur sometimes in service.

$$P(\overline{A}\,\overline{U}_L\overline{X}_M T_i V_i) = 1 \cdot 0.99 \cdot 0.05 \cdot 0.2 \cdot P(\overline{X}_M) \approx 10^{-2} \cdot P(\overline{X}_M) \leq 10^{-5}$$

$P(\overline{X}_M) \leq 10^{-3} \Leftarrow$ So for this situation we have a limit for the occurrence of manufacturing violations.

This type of loss of structural integrity clearly sets limits on how much property reduction can be considered passable in the 'practical world' and what is an unacceptable result of manufacturing process violation under all circumstances.

The reduction of structural properties due to manufacturing process failures is a complicated issue. It is dependent on material type, in particular how different types of property reductions interact. The process development for the material at hand must include a case-by-case basis determination of how process failures influence reductions of the following, and how they interact:

- strength/stiffness
- damage resistance
- damage growth rates
- damage tolerance
- manufacturing flaws
- fatigue resistance.

The whole issue related to manufacturing, assurance and control of quality and the protection against the worsening in service of manufacturing deficiencies needs a detail evaluation for each material system, process and structural concept. Practical limits and the interaction between reductions of properties controlling damage resistance, damage growth rates and damage tolerance and what is the dominating feature for the material system at hand can be solved through a practical focus.

Reductions in **stiffness properties**, regional effects in particular, can have bad effects on buckling and dynamic stability and are not, in general, detectable by inspection. However, spurious vibrations, reported when encountered, often can lead to an effective safety management under the auspices of 'operation'.

Damage resistance, **damage growth** and **damage tolerance** effects are also material system dependent, either compromised on the unit (PSE), level through process failures, or locally as in the case, for example, of local fiber waviness. Control or minimization of these effects needs to be dealt with on a case-by-case basis. The relative worsening of the three damage-associated properties is very important if avoidance is not to be the only practical approach. The results associated with probability control will be discussed in Chapter 3.

It is clear that a substantial reduction in damage resistance will only be manageable through a high-quality 'walk-around' inspection program. If at the same time both damage growth rates and damage tolerance are worsened, then only walk-around inspection and fail-safe design will lead to practical solutions. However, the opportunity exists to produce acceptable results through safety management for materials for which these three effects are demonstratively different.

We now will investigate the scenario for which we have both initial property reduction and accidental damage in service. Manufacturing process failures and accidental, service damage produce the following probability expression:

$$P(\bar{X}_{MO}Y_tD_{4t}D_{5T}\bar{B}_{LT}\bar{H}_T) = P(\bar{H}_T \mid D_{5T}) \cdot P(\bar{B}_{LT} \mid D_{5T}\bar{X}_MY_t) \cdot P(D_{5T} \mid D_{4t}Y_t\bar{X}_{MO})$$

$$\cdot P(D_{4t} \mid Y_t\bar{X}_{MO}) \cdot P(Y_t) \cdot P(\bar{X}_{MO})$$

$$\text{and } P(\bar{X}_{MO}) = \sum_{I=3}^{5} P(\bar{X}_MT_iV_i)$$

Taking into consideration estimates of the effect of property losses on the numerical results, the following results are shown for the total probability (RHS):

$$10^{-3} \cdot 0.5 \cdot 0.5 \cdot 0.5 \cdot 10^{-2} \cdot 0.3 \cdot 0.6 \cdot \text{p}_M = 2.3 \cdot 10^{-7} \cdot \text{p}_M < 8 \cdot 10^{-13}$$

This gives the following result:

$$\text{p}_M < 3.5 \cdot 10^{-6}$$

which would be a difficult value to achieve routinely for manufacturing process failures, (compare the results from Equation (2.13)), rendering walk-around inspections the only possibility.

Finally, the reduction of **fatigue resistance** is the only remaining property reduction described. The safety management needed for this item would focus on the performance of the structure towards the end of the service objective, by intensifying the inspections during that time. Further discussion is to be found in Chapter 3.

2.1.2 *Increase above limit load*

The contribution to this effect from manufacturing process failures is in general focused on installation of fasteners and assemblies, but in some cases local changes of stiffness properties can be a factor. In the majority of cases the following applies:

- Introduction of self-equilibrating load systems, e.g. fastener 'clamp-up';
- Local redistributions of internal load, e.g. spurious variations of fastener/hole fits;
- Variation of strength distributions in joints due to geometry mistakes, e.g. fastener edge margin violations, or inadequate torque.

These effects all result in local variations of load levels for limit conditions and will be illustrated below in Example 2.5, in which a typical situation will be described.

Example 2.5 We assume normally distributed strength, and B-values for design. The first case has a 30% local increase:

$$\text{Probability of violated ultimate integrity is: } \Phi\left(\frac{1.30LLR - \mu_x}{C_V\mu_x}\right)$$

B-value practice yields for $C_V = 0.10$: $\mu_x = 1.13LLR$
The value of the parenthesis above becomes: $t = 1.50 \Rightarrow \Phi(1.50) = 0.93$
And for a 20% increase we find: $\Phi(0.6) = 0.73$

An assumption of both normally distributed strength and local internal loads yields a probability of failure, p_f, for:

$$\text{a 30\% load increase: } p_f = 0.50$$
$$\text{a 20\% load increase: } p_f = 0.17$$

This underscores the importance of an effective safety management that recognizes the local nature of these kinds of effect, and whereas in the 'metal world' premature fatigue damage reveals the problem, this may not be the case in the 'composite world'. Again, avoidance of these kinds of violation level by meticulous quality assurance and control seems to be 'the answer'. This situation describes the effect of both added load systems and local redistributions.

2.1.3 *Spurious and/or unreported damage*

This section describes consequences of mechanical damage inflicted during the different manufacturing processes and includes de-laminations, de-bonds, inclusions, gages, accidental damage such as tool drops, moves on the factory floor, transportation, etc. – all the types of damage that constitute a threat to safety and are not detectable before the vehicle enters service. The serious nature of these flaws includes the facts that they may be inflicted in inaccessible airplane locations and that they often involve large-scale internal damage without any visible, external signs, and are therefore difficult to detect.

These types of damage constitute threats to:

- ultimate integrity
- fail-safe integrity
- limit integrity.

Therefore, requirements must be very well understood for each individual case. Detail design becomes the primary consideration. Mixtures of materials, different types of material and different designs all require different considerations, but the important point is that routine solutions will have to be replaced by informed engineering. It is clearly unacceptable to enter service with lost limit integrity. Lost ultimate integrity causes loss of fail-safe integrity and therefore corrupts limit integrity; so, when fail-safety is the primary damage-tolerance weapon, detection is the prime guardian of integrity. The detail design must be approached from the standpoint of the effects of manufacturing flaws and their detection.

Detection of manufacturing flaws in quality-control processes and in scheduled inspections is a very important subject for the monitoring and updating of 'safety levels' and a prime target for the design learning process.

SUMMARY

The safety threats originating in failed manufacturing processes are:

- reduced structural properties (e.g. reduced strength);
- aberrant geometry (e.g. inadequate edge margin);
- additional internal loads (e.g. clamp-up);
- undetected flaws (quality-control failure).

All can be serious threats to safety and are often difficult to detect, hence the need to use fail-safe designs.

2.2 INTEGRITY AND MAINTENANCE

The important safety features to be avoided in conjunction with this safety element are:

- mistakes in scheduled maintenance (e.g. removal of structural units);
- 'sloppy' inspections (damages remain undetected);
- repairs lacking in structural integrity.

The mistakes listed above often come in several versions. The first could cause abnormal wear that reduces strength or creates the need for excessive activation forces, both of which could affect loss of integrity. The second is sensitive to violations of specified inspection procedures, resulting in large-scale damage remaining undetected in service and loss of integrity. The third results in violated ultimate integrity and, consequently, loss of fail-safe integrity and violation of damage tolerance requirements under certain circumstances.

The total threat includes reduced structural properties and a local increase in internal load levels, and clearly is sensitive to the detail design of the structure being considered. The general protection against these types of safety threats involves establishing meticulous quality assurance and quality control for the processes involved, including design reviews and compliance demonstrations, and less reliance on standards as the field has strong, case-to-case dependences.

SUMMARY

Safety improvements must emerge from the knowledge that structural integrity is being threatened by reduced structural properties, unexpected damage and increases in the local, internal loads.

The main safety threats to integrity due to process violation in maintenance are:

- reduced structural properties;
- violated limit load values;
- inadequate detection of large-scale damage.

Safety management is dependent on detail design, structural concept and material mixture and, necessarily, is an integral part of structural design.

2.3 INTEGRITY AND OPERATION

The safety threats potentially associated with operation are:

- exceeding maximum design loads in service (excursions beyond limit load);
- operational exposure to severe environments (e.g. hail impact in flight);
- not reporting structural damage (e.g. collision with ground vehicles);
- poor 'walk-around' inspections.

The effects of these threats involve:

- excursions beyond limit loads;

- exposure to damage of varied size, temperature, pressure and chemicals outside the legal design range or duration, resulting in excessive reductions in structural property;
- loss of limit integrity due to unreported large damage.

The main tools of safety management, in this context, are education and training of personnel and strengthening of reporting procedures, including avoiding the punishment of the person reporting the 'problem'.

2.4   INTEGRITY AND REQUIREMENTS

To satisfy the objective safety levels, the requirements (regulations and criteria) must be without ambiguity. Present regulations (FAR 25) have a sweeping definition: 'Limit load is the largest load expected in service', is nebulous and therefore difficult to interpret. Present practices in damage tolerance apply the following definition: 'Limit load is the largest load reached within the period of service objective'. One could ask what the sources of the resulting vagueness are. One of the answers is that focusing on internal loads leads to a result different from that arrived at using external loads. Another is that external loads come in many categories – static loads, dynamic loads, emergency loads, handling loads, ground loads, etc.

One might also ask why the definition is not based on the 'internal, positive and negative load that will happen with a probability of 1'. This definition could be coupled with an individual load-case analysis that recognizes that static-load cases have a single peak during a flight while dynamic-load cases can have a series of peaks with about the same value. Crash pressures, for example, are of a complete different nature, and certainly are not under pilot control. Maximum (limit cabin-pressure) pressure is very tightly coupled with the reliability of the air-handling system. The differences need to be considered. It would be a large step toward avoiding excursions beyond limit load.

One would also expect that, at a time when safety in service needs to improve to respond to increasing traffic volumes, regulations and criteria would include required safety levels, or procedures for controlling safety levels in service and updating safety-preserving procedures.

It would also seem necessary to have damage-size criteria or combined detectability and size requirements for limit integrity.

Finally it is desirable that we have a 'crisp' definition of the design environments, (temperatures), for both ultimate and limit, critical designs, or at least procedures for how to establish the requirements.

The resulting regulatory requirements for composite structural design are:

- a precise and realistic limit load definition;
- a combined damage size and detection requirement for limit design;

- a definition of required design requirements for environmental effects;
- a special process for establishing hybrid (composites and metals) design criteria.

2.5 INTEGRITY SUMMARY

Safety management can be identified in terms of avoidance of unsafe events affecting safe designs. Table 2.1 describes the potentially spurious effects from different elements of safety. The effects are ordered as:

1. reduced properties;
2. added (internal) loads;
3. damage (inflicted in process);
4. exceeded limit load;
5. exposure to extreme environments.

The elements-of-safety process failures, potentially being the source of these effects, are also identified in Table 2.1. Table 2.2 describes the relation between the identified effects and three different structural integrities.

Table 2.1 Elements versus effect

	Effects	Elements				
#	Description	Manuf.	Maint.	Oper.	Req.	Comments
1	Red. propert.	•	•	–	–	For life
2	Added loads	•	•	–	–	For life
3	Damage	•	•	•	•	Long periods
4	Exceed. lim.	–	–	•	•	Temporary
5	Extr. env.	–	–	•	•	Temporary

Table 2.2 Integrity and effects

	Effects					
Integrity	Red. prop. 1	Add. loads 2	Damage 3	Exc. limit 4	Extr. env. 5	Comments
U_L (limit)	• (11–15)	• (21–23)	• (15)	• (31, 41)	• (32, 43)	
U_U (ult.)	• (11,!4)	–	–	–	• (43)	
U_F (fail-safe)	• (11)	• (23)	• (15, 33, 34)	• (31, 41)	• (11, 32, 43)	

Table 2.2 includes the effects, a detailed breakdown of the properties involved, and the specific relation to different integrities (see numbers in parentheses).

The three integrities identified play different roles from a safety standpoint, and we will start looking into the relation to safety management by investigating limit integrity, U_L. Table 2.2 lists the effects threatening integrity in terms of different elements with the numbers shown in each column. The typical expression for the probability of loss of integrity is of the form:

$$P(\bar{U}_L) = \sum_{k=1}^{4} P(\bar{U}_{Lk}) \tag{2.14}$$

and each term can be expressed as:

$$P(\bar{U}_{L1}) = \sum_{j=1}^{5} P(\bar{U}_{L1j})$$

and a typical term is:

$$P(\bar{U}_{L1j}) = P(\bar{B}_{L1j}\bar{X}_M T_{1j} V_{lj}) = P(\bar{B}_{L1j} \mid \bar{X}_M T_{1j} V_{lj}) \cdot P(V_{lj} \mid T_{lj}\bar{X}_M)$$
$$\cdot P(T_{lj} \mid \bar{X}_M) \cdot P(\bar{X}_M)$$

The next chapter investigates and describes the structural threats in terms of probabilities and what could be tolerable and what should be avoided. The reduction of 'fatigue resistance' due to manufacturing processes failures will be the subject of a separate study, because it is not anticipated to become a problem until late in service.

Example 2.6   We assume a normally distributed strength, and a manufacturing process failure that only affects strength; we evaluate the typical term above for limit strength. The result is:

$$P(\bar{U}_{L11}) = \Phi \cdot k \cdot 0.2 \cdot 10^{-4}$$

for 30% reduction

$$P(\bar{U}_{L11}) = 0.96 \cdot 10^{-1} \cdot 0.2 \cdot 10^{-4} = 2 \cdot 10^{-6}$$

for 10% reduction

$$P(\bar{U}_{L11}) = 10^{-4} \cdot 0.3 \cdot 0.2 \cdot 10^{-4} = 0.6 \cdot 10^{-9}$$

Where the evaluation of the probability of falling below the allowable value when B-values are used, Φ, for limit is

For B-value

$$\Phi(t) = \Phi\left(\frac{x - \mu}{C_V \cdot \mu}\right) = 0.10 \Rightarrow t = -1.3 \Rightarrow x = 0.87\mu$$

30% reduction

$$\Phi\left(\frac{\dfrac{0.87\mu}{1.5} - 0.7\mu}{0.1 \cdot 0.7\mu}\right) = 0.96$$

10% reduction

$$\Phi\left(\frac{\dfrac{0.87}{1.5 \cdot 0.9}}{0.1}\right) = 10^{-4}$$

The numerical evaluation of $P(\bar{U}_L)$ shows that pursuing this approach leads to a limit for passable range and for an upper bound of the structural property reduction that defects in the manufacturing processes can be allowed to produce. It is interesting to note that the probability of manufacturing process failures in general have a large influence on the resulting safety standards.

2.6 CONCLUSIONS

The effects of all defects of the processes of all elements of safety can be interpreted as:

- reduction in structural properties;
- exceedance of limit load;
- spurious and/or unreported damage.

Effective inspection, timely detection and reporting have a market effect on protecting and restoring level of safety.

Ultimate, limit, fail-safe integrity and damage detection are all parts of structural safety.

Chapter 3 provides an all-around assessment of how the probabilities of the additional elements of safety influence structural safety and the control of the level of safety.

Manufacturing process failures are very important for structural safety because they very often occur without detectable flaws, and they very often affect several defective, adjacent load-paths (e.g. on a wing or stabilizer surface), which causes loss of fail-safe integrity.

2.7 THE PROCESS QUALITY OF THE ELEMENTS OF SAFETY

The tendency of striving toward a 6-sigma process quality, for both the requirements and the application of the process to composite structure, is laudable. To implement and improve all elements of safety would serve the quest for safety. The previous discussion in Chapter 1 has shown the huge impact manufacturing defects have on safety. It seems that successful composite structural innovation with modern material integration into transport-category airplanes will depend on adaptation to and integration of all elements of safety.

The basic consequence of properly exploiting a 6-sigma process quality is a probability of a defect inflicted by the process of $3.4 \cdot 10^{-6}$. It would be desirable that a definition of defects included non-matching, albeit allowable, strengths of composite structure. (For example, B-values, which for a sampling test of n specimens is described in the following example.)

Example 2.7   We assume a result of 20 successes ($>$ B-values) with 6-sigma; the probability is

$$\Rightarrow (1 - 3.4 \cdot 10^{-6})^{20} = 0.9993$$

and by the definition of B-value: $0.90^{20} = 0.12$. Using mean as design value: $0.5^{20} = 9.5 \cdot 0^{-7}$.

The majority of structural data comes with qualities used in Example 2.7, which begs the question of how one avoids including positive values among the defects of process sampling.

Chapter 3
Elements of Safety and Design Data

Chapter 2 discussed the influence of process failures on the probability of loss of integrity. This chapter will extend the discussion to the probability of an unsafe state. Table 2.1 provides data on the nature of the effects of the process failure. The reduction of fatigue resistance was identified as an important factor toward the end of service. Increases of damage frequency can point to further root causes. In addition, we have identified effects that can last through the entire period of service, or for a limited period, or which are short-term events taking place during a single flight. We now will investigate the influence of time on the loss of integrity – in this case limit integrity – and initially the events that do not involve mechanical flaws.

The probability of preserved limit integrity at 'roll-out', O, can be written as:

At 'O', 'roll-out'

$$P(U_{LO}) = P(U_{LDO} \cdot U_{LMO} \cdot U_{LRO}) = P(U_{LDO}) \cdot P(U_{LMO}) \cdot P(U_{LRO})$$
$$= [1 - P(\bar{U}_{LDO})] \cdot [1 - P(\bar{U}_{LMO})] \cdot [1 - P\bar{U}_{LRO})]$$
$$\Rightarrow \text{ for lost limit integrity at 'roll-out'}$$
$$P(\bar{U}_{LO}) \approx P(\bar{U}_{LDO}) + \bar{P}(\bar{U}_{LMO}) + P(\bar{U}_{LRO}) \tag{3.1}$$

For lost integrity at T_1:

$$P(\bar{U}_{L1}) \approx P(\bar{U}_{LD1}) + P(\bar{U}_{LM1}) + P(U_{LI1}) + P(\bar{U}_{LO1}) + P(\bar{U}_{LR1}) \tag{3.2}$$

Service involves a mix of initially possible violations and potential mishaps in maintenance and operations.

Chapter 2 identified a number of effects that can be grouped in two categories, decreased properties affecting, for example, strength, and violations that increase internal loads beyond design loads. The relation between mechanical damage and the success of elements of safety renders the subject complicated, the two equations that control safe processes are given below.

The probability of the joint event describing damage and preserved limit integrity is:

$$P(\bar{X}_t U_{Lt}) = P(\bar{X}_t U_{LDt} U_{LMt} U_{LIt} U_{LOt} U_{LRt}) \tag{3.3}$$

and the probability of mechanical damage when all processes are violated:

$$P(\bar{X}_t \bar{U}_{Lt}) = P[\bar{X}_t (\bar{U}_{LDt} \cup \bar{U}_{LMt} \cup \bar{U}_{LIt} \cup \bar{U}_{LOt} \cup \bar{U}_{LRt})] \tag{3.4}$$

The detailed description between cause and effect and the timing of different violations will now be studied in combination with 'mechanical damage'. The dominating concern in safety management is described by Equation (3.4) for both unique and combined violations. Equation (3.5) describes the safety problem:

$$P(\overline{A}\,\overline{U}_L) = P(\overline{A} \mid \overline{U}_L) \cdot P(\overline{U}_L) \tag{3.5}$$

Here the right-hand-side product of 'probability of failure', given loss of limit integrity and the probability of loss of limit integrity, sets both the standard and the limit in protecting against failure. The significance of mechanical damage in safety management will be discussed in the following sections.

The probability of an unsafe state for the structure constituting a principal structural element (PSE) has contributions from: design, manufacturing, maintenance, operation and requirements. Their individual contributions vary with the case-to-case situation and considerable uncertainty is involved. The evolution of the reduced uncertainty comes from the monitoring and updating of service data, and ultimately from the long-term understanding and knowledge associated with the specific material, process and structural concepts involved.

Backman (2005) contains an extensive study of design, and this chapter will focus on the threats from the other four elements of safety. The exception is mechanical damage, which will be included in design and treated as a total threat no matter what the cause is.

3.1   MISHAPS IN MANUFACTURING

The effects of manufacturing mishaps can be listed as:

1.   Reduced structural properties (e.g. process failures);
2.   Added internal loads (e.g. 'clamp-up');
3.   Incorrect geometry (e.g. mis-located fasteners);
4.   Accidental damage (e.g. de-bonds, inclusions).

Here process control, quality management and quality control are the weapons for optimum quality at entry into service.

3.1.1   Reduced structural properties

Reduced structural properties due to process failures are difficult to detect, especially when mechanical, detectable damage is not present. The reduced properties must be evaluated on a case-by-case basis. The effects can be:

- reduced strength/reduced stiffness;
- decreased damage resistance;

- increased damage growth rates;
- decreased damage tolerance;
- manufacturing flaws causing threats to fail-safe integrity;
- reduced fatigue resistance (increased accumulation of damage).

Reduced strength and/or stiffness can cause a loss of limit integrity, and the probability can be expressed as

$$P(\bar{U}_{LM}) = P(\bar{B}_L \bar{X}_{M11})$$

where $\bar{X}_{M11}$ is manufacturing caused reduced strength.

The probability of a reduced property without mechanical damage is, for example, for strength:

$$P(\bar{U}_{LM}) = P(\bar{B}_L \bar{X}_{M11} X) = P(\bar{B}_L \bar{X}_M T_1 V_1 X) = P(\bar{B}_L \mid \bar{X}_M T_1 V_1 X)$$
$$\cdot P(X \mid \bar{X}_M T_1 V_1) \cdot P(V_1 \mid T_1 \bar{X}_M) \cdot P(T_1 \mid \bar{X}_M) \cdot P(\bar{X}_M) \qquad (3.6)$$

Here:

$\bar{U}_{LM}$:   loss of limit, 'L', integrity due to manufacturing mishap
$\bar{B}_L$:   limit residual strength is less than limit load requirement, LLR
$\bar{X}_{M11}$:   manufacturing mishap (reduced strength, (11))
X :   no mechanical damage
T_1 :   mishap Type '1', reduced strength due to manufacturing error
V_1 :   value of resulting effect.

Equation (3.6) applies to limit strength reduction due to process failure in manufacturing. The following example investigates numerical consequences in the context of setting requirements.

Example 3.1 Figure 3.1 illustrates strength reduction by a factor of 'k', which is assumed to apply to the mean of normal, probability, strength distributions.

We assume that the B-values are used for ultimate, that $C_V = 0.10$ yields the probability of being below limit value, and that damage is contained in the ultimate range:

$$\Phi\left(\frac{\dfrac{0.87\mu}{1.5} - k\mu}{C_V k\mu}\right) = \Phi\left(\frac{5.8}{k} - 10\right)$$

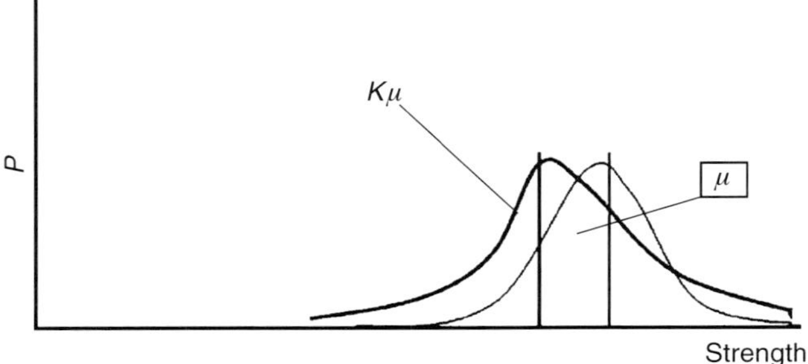

Figure 3.1 Reduced properties.

k	t	$\Phi(t)$
0.80	-2.75	0.003
0.90	-3.56	$1.6 \cdot 10^{-4}$

The probability of loss of limit integrity due to reduced strength can be evaluated by Equation (3.6). The equation can be written as

$$p_{\min} = P(\bar{B}_L \mid \bar{X}_M T_1 V_1 X) \cdot P(X \mid \bar{X}_M T_1 V_1) \cdot P(V_1 \mid T_1 \bar{X}_M) \cdot P(T_1 \mid \bar{X}_M) \cdot P(\bar{X}_M)$$
$$= \Phi(t) \cdot P(X \mid \bar{X}_M T_1 V_1) \cdot p_T \cdot 0.5 \cdot 10^{-4} = \Phi(t) \cdot p_{\text{sub}}$$

k	$\Phi(t)$	p_{sub} (per Eq. 3.6)	$p_{\min}$
0.8	0.003	$0.5 \cdot 10^{-5}$	$1.5 \cdot 10^{-8}$
0.9	$1.6 \cdot 10^{-4}$	10^{-5}	$1.6 \cdot 10^{-9}$

This numerical evaluation shows that only marginal reductions in strength produce acceptable limit integrity. We will now investigate other types of mishaps and evaluate required quality assurance and damage control limits. Reduced stiffness produces an analogous situation to reduced strength.

A similar situation that includes initial, mechanical damage can be expressed as

$$P(\bar{U}_{LM}) = P(\bar{B}_{LM} \bar{X}_{M11} \bar{X}_O D_{iO} \bar{H}_O) = P(\bar{B}_{LM} \mid \bar{X}_{M11} \bar{X}_O D_{iO} \bar{H}_O) \cdot P(\bar{H}_O \mid D_{iO})$$
$$\cdot P(D_{iO} \mid \bar{X}_O \bar{X}_{M11}) \cdot P(\bar{X}_O \mid \bar{X}_{M11}) \cdot P(\bar{X}_{M11}) \tag{3.7}$$

Here:

$\bar{H}_O$: not detected by quality control
D_{iO} : damage size interval at roll-out.

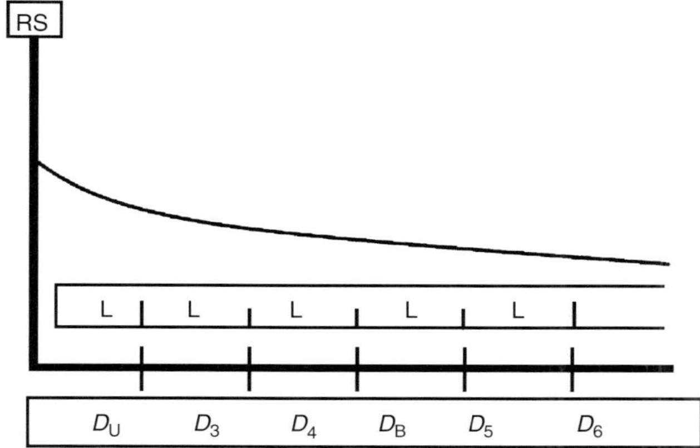

Figure 3.2 Damage interval and sizes.

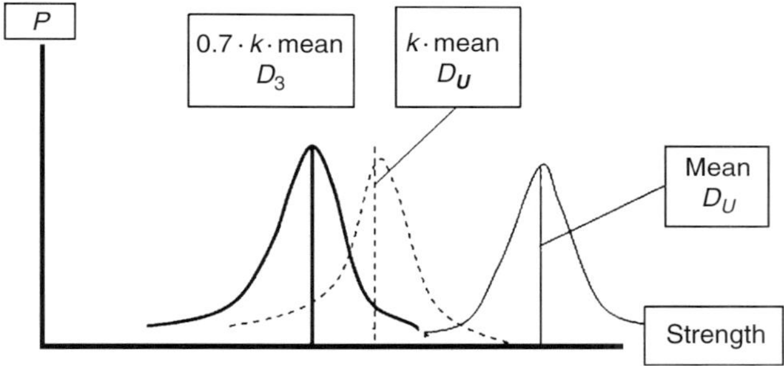

Figure 3.3 Strength distributions for reduction and damage.

Example 3.2　This example is an extension of reduced strength to also include the effects of mechanical damage. Equation (3.7) describes the events involved and their values at different levels of strength reduction. The effect of an increase in damage size is assumed to obey the square root of the size ratio (Figures 3.2 and 3.3).

A change from D_U to D_3 yields ~ 0.7.

The following terminology is used in the table below:

The following variables are included:

k	:	strength reduction
$\Phi(t)$	:	allowable value quality
i	:	damage interval index
$\Phi(t_1)$	:	RS allowable quality corrected for larger damage
p_1	:	$P(\bar{B}_{LM} \mid \bar{X}_{M11}\bar{X}_O D_{io}\bar{H}_O)$

$$p_2 \;:\; P(\bar{H}_O \mid D_{iO})$$
$$p_3 \;:\; P(D_{iO} \mid \bar{X}_O \bar{X}_{M11})$$
$$p_4 \;:\; P(\bar{X}_O \mid \bar{X}_{M11})$$
$$p_5 \;:\; P(\bar{X}_{M11}) = P(\bar{X}_M T_I V_I) = P(V_I \mid T_I \bar{X}_M) \cdot P(T_I \mid \bar{X}_M) \cdot P(\bar{X}_M)$$

k	$\Phi(t)$	i	$\Phi(t_1)$	p_2	p_3	p_4	P_5	p_{min}
0.8	0.003	3	–	10^{-1}	0.5	0.5	$0.4 \cdot 0.5 \cdot 10^{-4}$	$1.2 \cdot 10^{-9}$
0.8	–	4	0.63	10^{-2}	10^{-2}	0.5	$0.2 \cdot 10^{-4}$	$0.63 \cdot 10^{-9}$
0.9	$1.6 \cdot 10^{-4}$	3	–	10^{-1}	0.5	0.5	$0.2 \cdot 10^{-4}$	$0.8 \cdot 10^{-10}$
0.9	–	4	0.21	10^{-2}	10^{-2}	0.5	$2 \cdot 10^{-4}$	$0.21 \cdot 10^{-9}$

It is interesting to note that the presence of damage sizes that exceed the ultimate range when complemented by good quality control, for d_S not belonging to D_3, can be favorable compared to the situation in the ultimate damage range. It also is clear from the numerical results that enforcing limits for strength reductions and damage limits needed to control the effects of manufacturing mishaps is a case-by-case endeavor with emphasis on detailed material behavior.

It is important to accept (1) the case-by-case situation in determining how different properties degrade due to the same process defects; (2) that process quality requirements change; and (3) that interactions between different types of defects change for different types of composite structure.

3.1.2 Added internal loads

The result of many types of manufacturing mishap is induced internal loads: self-equilibrating loads as encountered due to, for example, inadequate shimming and accompanying 'clamp-up' in mechanical joints, or local redistribution in load transfer regions because of spurious fastener flexibilities for example on hole and fastener tolerances, or varying 'torque', or misplaced fasteners or even unintended stiffness variations in bonded joints. Figure 3.4 illustrates the resulting impact on limit integrity and associated probabilities. It shows a situation for which strength is unchanged, the structure sized for limit load and that a manufacturing mishap increases maximum load by a factor of 'k.'

Example 3.3   This example assumes a normal strength distribution, a B-value that is equal to LLR and 0.87μ respectively and $C_V = 0.10$. Figure 3.4 illustrates a manufacturing mishap that increases the internal load where by a factor of 'k.' We now ask what the probability of violating the strength at limit is:

The ultimate B-value assumption (B-value is $0.87 \cdot 1.5 \cdot$ LLR), leads to:

$$P(\bar{B}_{LM}) = \Phi\left(\frac{kLLR - 1.15 \cdot LLR \cdot 1.5}{\sigma}\right) = \Phi\left(\frac{0.58k - 1}{0.1}\right)$$

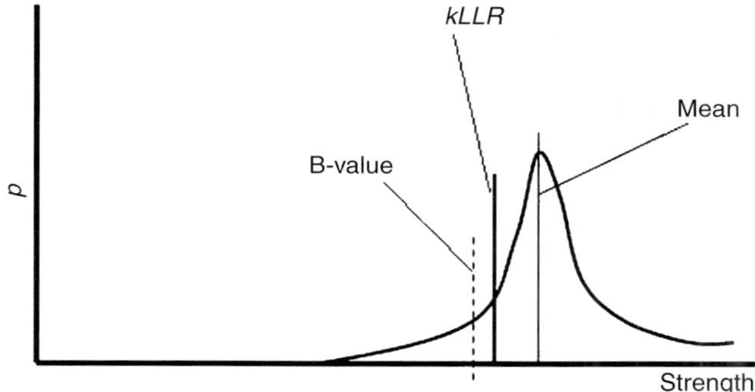

Figure 3.4 Increased load beyond limit.

k	t	$\Phi(t)$
1.2	-3.0	0.0014
1.3	-2.46	0.0070
1.4	-1.88	0.0300

It is clear that this range of load increases constitutes a serious threat.

3.1.3 *Reduced damage resistance, increased damage growth rate and reduced damage tolerance*

Two different situations arise, one involving the coincident compromise of all three structural properties, the other being when these properties are influenced by independent mishaps.

A very complicated situation exists between:

- damage resistance reduction;
- increased damage growth rates;
- damage tolerance.

They all seem often to show similar responses to manufacturing mishaps, so we start with the situation in which all three effects are created by one type of mishap.

Example 3.4 The case selected is the most serious and can be represented by the following equation:

$$P(\bar{U}_{LM}) = P(\bar{B}_{LT} \mid D_{5T}\bar{X}_M) \cdot P(\bar{H}_T \mid D_{5T}) \cdot P(D_{5T} \mid \bar{X}_M \bar{X}_0 D_{40}TV)$$
$$\cdot P(D_{40} \mid \bar{X}_M \bar{X}_0 TV) \cdot P(V \mid T\bar{X}_M) \cdot P(T \mid \bar{X}_M) \cdot P(\bar{X}_M)$$

A reduction in structural performance of 20% results in an example probability of

$$1 - P(U_{LM}) = 0.63 \cdot 10^{-3} \cdot 0.18/0.8 \cdot 10^{-2} \cdot 0.2^2 \cdot 10^{-4} = 4 \cdot 10^{-11}$$

These ranges of value place these types of mishap low on the list of manufacturing threats for materials and processes that support this behavior and reinforces the fact that a case-by-case evaluation is important in the determination of requirements. It is interesting to note that the '6-sigma' approach from quality management ideas seems to provide a convincing starting point for quality assurance for composites processes.

A look at the separate effects sheds some additional light on requirements. We start with damage resistance and the requirement that the initial damage in the range of region D_5 is not acceptable except when the loaded structure is fail-safe. We assume that damage resistance controls initial damage according to the equation:

$$D_S = \frac{E \cdot I}{R}$$

a reduced damage resistance by a factor of 'k' $\Rightarrow$ a new damage size of $\dfrac{D_S}{k}$
Here:

E : constant energy level
I : impact characteristic
R : damage resistance.

We will now investigate $k = 0.80$, and we will assume that the distribution is normal.

Thus the only design solution is the fail-safe approach mentioned above; otherwise, requirements for quality assurance must be implemented that are significantly more stringent. The only other protection would be walk-around inspection if the PSE is accessible.

A similar reduction in 'damage tolerance' would result in an 'Unsafe State'.

3.1.4 *Reduced fatigue resistance*
This refers to a term from the 'metal world' that represents the main cause of widespread mechanical damage. The 'composites world' presents more diverse situations, for which there are three major causes of intensified mechanical damage. These are:

- reduced fatigue resistance;
- increased damage growth rates;
- increased accidental damage accumulation.

Hence the threat of widespread fatigue in the 'metal world' is, in the 'composites and hybrid worlds' equivalent to the combined effects of the above three causes.

The typical equations for these three causes are:

$$P(\bar{B}_L) = P(D_f \bar{X}_{ft} \bar{H}_T) = P(\bar{H} \mid D_f \bar{X}_{ft}) \cdot P(D_f \mid \bar{X}_{ft}) \cdot P(\bar{X}_{ft}) \qquad (3.8)$$

where for accidental damage index 'f' is replaced by 'a' and for damage growth index 'f' is replaced by 'u'.

Example 3.5 This example is a numerical illustration of widespread accumulation of damage. We assume that we are dealing with a PSE of 50 DDPs (Detail Design Points). We also assume that when four adjacent DDPs are damaged, limit integrity is lost.

$$\text{Fatigue damage: } P(\bar{X}_{Lf}) = {\sim}1 \cdot 10^{-4} = 10^{-4}$$
$$\text{Accidental damage: } P(\bar{X}_{la}) = 10^{-2}$$
$$\text{Damage Growth: } P(\bar{X}_{Lu}) = 10^{-2} \cdot 0.5 \cdot 10^{-1} - 0.5 \cdot 10^{-3}$$
$$\sum \approx 1.06 \cdot 10^{-3}$$

We assume a rectangular pattern of DDP's (five rows of 10 points each), yielding:

$$N = 6 \cdot 5 \Rightarrow P(\bar{A}) = 30 \cdot [1.06 \cdot 10^{-3}]^4 = 3.8 \cdot 10^{-11}$$

This result clearly illustrates the need to study the system at hand to determine the relative criticality, which is particularly relevant to determining the approach to inspection late in the life of the vehicle.

From this numerical cavalcade emerge illustrations of different types of process mishaps, and the complications encountered when clearly detectable damage is not involved. These complications often require additional attention to quality assurance to render the events improbable, introducing approaches in detail design that make walk-around inspections effective, or the establishing of fail-safe detail designs that can survive load-path failures at limit load, especially in compression, a failure that is much more violent than the often gentle tension failures encountered under fatigue loads in metallic structures.

3.2 QUALITY ASSURANCE

The Six Sigma concept embraced in quality management circles often utilized in quality assurance and process developments in composites. The term comes with many controversial aspects. The normal distributions of process measures are postulated with the assurance that many processes are not normally distributed. The term is very ambiguous as it implies that process limits should be contained within six standard deviations. However, standard practices assume a 1.5 sigma 'drift' and the often-quoted value

3.4 defects per one million opportunities

is based on the normal assumption, i.e. the $1.5 \cdot \sigma$ 'drift', and on a one-sided limit. The probability of results outside process limit, Q, is written as:

$$P(Q) = \Phi\left(\frac{\mu - 4.5\sigma - \mu}{\sigma}\right) = \Phi(-4.5) = 3.4 \cdot 10^{-6} \tag{3.9}$$

This equation describes the probability that the process will produce the results shown below, a lower bound, LB, set as:

$$\text{LB} = \mu - 4.5 \cdot \sigma$$

The demonstrated 'drift' in processes adds a troublesome uncertainty to most manufacturing 'steps' because it produces more defects in time and makes the monitoring/updating cycle for preservation of safety levels more complicated, in this case becoming another aging phenomenon. However, the elimination of defects in an organized way that includes a gradual accumulation of defects in the interval 6σ to 7σ of 0 to 10% will establish a long-term goal of no defects below 6σ.

If we identify and classify the processes that constitute manufacturing, the possible results would incorporate some variations of the following list:

- laminate (sandwich), processing
- part fabrication
- sub-assembly
- assembly
- final assembly
- final inspection
- compliance demonstration (testing).

The relative 'weight' of the manufacturing element in assigning structural safety requirements could be based on

$$P(\bar{Q}_M) = 7 \cdot 3.4 \cdot 10^{-6} = 2.4 \cdot 10^{-5} \tag{3.10}$$

The following sections contain investigations of the relative influences of the different elements of safety with the objective of establishing individual goals for all of them.

3.3   QUALITY CONTROL

A 'top-down' derivation of the requirements for probability of an 'unsafe flight' includes the following levels:

For the airplane: 10^{-5}
Improvement objective: 10^{-6}

Structural part: 10^{-7}

For PSE (50 assumed): $2 \cdot 10^{-9}$

If all DDPs are involved the design objective should apply to all.
The probability of an unsafe PSE can be written as:

$$P(\overline{S}) = P(\overline{S}_D) + P(\overline{S}_M) + P(\overline{S}_I) + P(\overline{S}_O) + P(\overline{S}_R) \leq 2 \cdot 10^{-9} \tag{3.11}$$

A rational safety goal is that the major effects are of roughly the same order of magnitude. Particularly if one or a few elements dominate, dramatic safety improvements can be achieved with relatively small effort and costs and with large pay-offs.

Backman (2005) demonstrates how the maximum value for the probability of an unsafe flight due to design inadequacies, $\overline{S}_D$, is:

$$P(\overline{S}_D) = P(\overline{B}_{LT}\overline{X}_\tau\overline{H}_\tau D_{5T}\overline{H}_T) \leq 10^{-10} \tag{3.12}$$

$\overline{B}_{LT}$: loss of limit integrity at 'T'
$\overline{X}_\tau$: damage present at 'τ'
$\overline{H}_\tau$: damage not detected at 'τ'
D_{5T} : damage size in region '5'
$\overline{H}_T$: damage not detected at 'T'

To get a closer look at the numbers produced for manufacturing we use the definition

$$P(\overline{S}_M) = P(Q_M\overline{H}_{QC}) = P(\overline{Q}_M) \cdot P(\overline{H}_{QC}) \tag{3.13}$$

The desired value of Equation (3.13) will now be investigated. The example below uses Equation (3.12).

Example 3.6 A variant of Equation (3.12) and Equation (3.13) yields:

$$2.4 \cdot 10^{-5} \cdot P(\overline{H}_{QC}) \approx 0.4 \cdot 10^{-9} \Rightarrow P(\overline{H}_{QC}) \approx 0.2 \cdot 10^{-4}$$

So, either this puts extreme pressures on improvements of the 'detection business', or the quality assurance process must be monitored closely, and the 'drift' must be kept at a compromise value. The following table could be used to select the allowable drift.

Lower limit	Drift	t	$\Phi(t)$	$P(\overline{Q}_M)$
$\mu{-}6\sigma$	0	-6.0	10^{-9}	
$\mu{-}4.5\sigma$	1.5	-4.50	$0.3 \cdot 10^{-5}$	$2.1 \cdot 10^{-5}$
$\mu{-}4.75\sigma$	1.25	-4.75	10^{-6}	$7 \cdot 10^{-6}$
$\mu{-}5\sigma$	1.0	-5	$0.29 \cdot 10^{-6}$	$0.21 \cdot 10^{-6}$
$\mu{-}5.25\sigma$	0.75	-5.25	$0.2 \cdot 10^{-6}$	$1.4 \cdot 10^{-6}$
$\mu{-}5.5\sigma$	0.5	-5.50	$0.14 \cdot 10^{-6}$	10^{-6}

If we were to select a drift of 0.75 and a probability $P(\bar{H}_{QC}) = 10^{-3}$ we get from equation (3.13):

$$P(\bar{S}_M) = 0.4 \cdot 10^{-9} = P(\bar{Q}_M) \cdot 10^{-3} \Rightarrow P(\bar{Q}_M) = 0.4 \cdot 10^{-6}$$

the value falls between a 'drift' of 1.5 and 1.0.

The answer for manufacturing, in supporting the required safety level, is to reduce the process drift to the neighborhood of 1.25 and to develop quality-control processes that only miss defects outside limits with a probability of 10^{-3}.

Especially difficult circumstances exist for panel allowable values, for which avoiding the violating of limit integrity for composite panels requires a $C_V < 0.06$.

This chapter represents a process for safety target setting and each material, process and structural concept will have to be dealt with on a case-by-case basis. The following safety evaluation will deal with the remaining elements. The total result could be considered the initial value for a specific case developed for a special situation.

3.4 ERRORS IN MAINTENANCE

The following activities included in maintenance influence safety:

- scheduled maintenance
- major inspections
- repairs.

Errors with significant influence on safety include reduced structural properties due to defects in activities such as lubrication or the increase in internal loads due to abnormal wear and tear.

Major deviations in inspection procedures allowing significant damage to go undetected, or suspect signs not to be acted on, or inadequate training of inspectors resulting in reduced detection of safety threatening damage, are all part of the inspection process.

The inadequate structural integrity resulting from the repair process that we must protect against include damage inflicted in 'remaining structure' with a resultant strength decrease (structural property reduction), process violations that compromise structural properties and geometric mistakes that increase internal loads (such as like gaps and clamp-up).

3.4.1 Scheduled maintenance

We start from the original Six-Sigma view and ask the question: 'What is the probability that a "defect" in the scheduled maintenance process will violate limit integrity?' The important events involved lead to the following equation:

$$P(\bar{B}_L \bar{Q}_{11} X) = P(\bar{B}_L \mid \bar{Q}_{11} X) \cdot P(X \mid \bar{Q}_{11}) \cdot P(\bar{Q}_{11}) \tag{3.14}$$

$\overline{B}_L$: residual strength less than *LLR*

$\overline{Q}_{11}$: scheduled maintenance process outside process limits

X : no mechanical damage present.

Example 3.7 Equation (3.14) represents the loss of limit integrity for a PSE where all the DDP's are affected by the process defect, and the assumption applied is that the process characteristics satisfy Six-Sigma quality requirements. Thus the probability of loss of limit integrity, $P(\overline{U}_L)$, is:

$$P(\overline{U}_L) = 10^{-5} \cdot 10^{-1} \cdot 3.4 \cdot 10^{-6} = 3.4 \cdot 10^{-12}$$

This result can be considered as representative failures in scheduled maintenance that involve reduction of structural strength. When the result is an increase in internal loads, the effect is analogous to similar results in the manufacturing process discussed in Section 3.1.2.

3.4.2 *Major inspections*

The safety of composites in service depends on many factors, as has been seen in the earlier parts of this chapter. The structural design of composites often depends on relatively slow growth of damage and the critical state of safety is driven by large-scale damage, either in association with the damage that is defined as maximum for damage tolerance or the damage used in the fail-safe limit design. As lost limit integrity leads to failure in a very limited time frame, the detection and repair of large-scale damage in a safe, consistent manner is an unconditional requirement. One of the scenarios in the design of safe structure describes the probability of an unsafe state as (see Backman 2005):

$$P(\overline{S}_T) = P(\overline{H}_\tau \overline{U}_{LT} \overline{H}_T) = P(\overline{H}_\tau \overline{B}_{LT} \overline{X}_\tau D_{5T} \overline{H}_T)$$

The design objective derived from this equation is to minimize probability. The importance of a reliable inspection process can be illustrated by focusing on the events surrounding time 'T' while the process is producing a defect:

$$P(\overline{X}_{12} \overline{B}_{LT} \overline{X}_T D_5 \overline{H}_T) = P(\overline{B}_{LT} \mid \overline{X}_{12} \overline{X}_T D_5 \overline{H}_T) \cdot P(\overline{H}_T \mid \overline{X}_{12} \overline{X}_T D_5)$$
$$\cdot P(D_5 \mid \overline{X}_T \overline{X}_{13}) \cdot P(\overline{X}_T \mid \overline{X}_{12}) \cdot P(\overline{X}_{12}) \tag{3.15}$$

$\overline{X}_{12}$ = the inspection process in the defective range.

The influence of the inspection process will be illustrated in the next example.

Example 3.8 We assume that the limit residual strength requirement has been violated, that the damage is in the D_5 range, and we will use Equation (3.15)

$$P(S_{I2T}) = 10^{-3} \cdot 1 \cdot 10^{-1} \cdot 10^{-1} \cdot 3.4 \cdot 10^{-6} = 3.4 \cdot 10^{-11}$$

So, we conclude that for this situation, major inspection defects can cause contributions to the probability of an unsafe state of the order of magnitude of

$$\Delta P(\overline{S}_{IT}) \approx 3.4 \cdot 10^{-11}$$

Here the inspection process is of Six-Sigma quality.

We will now proceed to an illustration of the repair process so that the total effect of scheduled maintenance can be assessed.

3.4.3 Repairs

The repair criterion used in this context is:

Damage must be detected and repaired before limit integrity has been lost.

The mishaps that influence the quality of the repair process are:

- damage inflicted in the removal and preparation phase;
- process defects in application phase ('bonded patch applied' and structural integrity not restored);
- bolted repair fastener installation defective.

The events describing the mishap that is not detected during the first phase are governed by the equation

$$P(\overline{S}_{I31}) = P(\overline{B}_{LI3}\overline{X}_{I3}\overline{X}_0\overline{H}_0 D_5) = P(\overline{B}_{II3} \mid \overline{X}_{I3}\overline{X}_0\overline{H}_0 D_5) \cdot P(\overline{H}_0 \mid \overline{X}_{I3}\overline{X}_0 D_5)$$
$$\cdot P(D_5 \mid \overline{X}_0\overline{X}_{I3}) \cdot P(\overline{X}_0 \mid \overline{X}_{I3}) \cdot P(\overline{X}_{I3}) \tag{3.16}$$

$\overline{S}_{I31}$: unsafe state due to a mishap in the first stage of the repair process
$\overline{X}_{I3}$: defect in maintenance process due to mishap during repair.

The next example illustrates a numerical result due to a mishap described by equation (3.16) under the assumption that the process is controlled by Six-Sigma.

Example 3.9 The site has undetected mechanical damage in the 'remaining' and the probability of the joint event is:

$$P(\overline{S}_{I31}) = 10^{-1} \cdot 10^{-2} \cdot 10^{-2} \cdot 10^{-1} \cdot 3.4 \cdot 10^{-6} = 3.4 \cdot 10^{-12}$$

The first element of the repair process could require much attention and caution should be taken in avoiding this result, especially if a number of DDP's are involved.

The next mishap of concern is bonding process failure that causes loss of specific structural properties. We will start with strength focus and describe the event guiding lost limit integrity and no damage being present. The probability of an unsafe state due to this joint event is described in principle in Example 3.1, where it is clear that even moderate strength reductions (0.8–0.9) can result in significant threats when compared with Example 3.9.

The remaining threat is due to defects in mechanical fastener installations, when the structure and the repair are designed for open-hole compression and filled-hole tension. This leaves varying bolt installation quality, resulting in overload in one or more fastener(s) as the main threat. Example 3.3 and Figure 3.4 illustrate the situation, and the consequences on the safety level are repeated. Only modest 'added loads' can be tolerated without serious structural damage being inflicted.

3.5   FAILURES IN OPERATION

The organizations responsible for safety promotion and preservation in service are participants of this element of safety, especially the following:

- operating the airplane so that limit loads are not exceeded;
- containing operation of flights inside the envelope for temperatures ($<T_{MAX}$);
- avoiding severe weather extremes, such as excessive wind gusts, thunderstorms and hailstorms;
- reports on ground damage and initiative taken for repairs;
- conduct walk-around inspections.

This is a particularly important list, because it itemizes part of the routine for every flight and describes actions that could save airplanes from disastrous failures by timely and informed action being taken during service.

3.5.1   *Exceeding limit loads*

Section 3.1.2 provides an investigation of manufacturing flaws that result in additional internal loads and which consequently make limit load violation more likely. The results of operational mishaps are more immediate because the actions taken are the direct cause of the limit load exceedance, the most likely remedy lies with improving the tolerances of the process that is intended to adjust load levels to safe values defined by the following:

Limit loads are the highest loads *expected* in service.

This applies to transport category vehicles. Returning to design decisions taken during development: these are made assuming safe operation, so the loss of limit integrity is caused by the interaction between, for example, limit level acceptable mechanical damage sizes and excessive load. The following equation and events apply:

$$P(\bar{U}_{LO}) = P(\bar{X}_t D_{5t} B_{Lt} \overline{HX}_{OT} T_{1T} V_T \bar{B}_{LT})$$

(3.17)

The participating events are:

$P(\bar{U}_{LO})$: loss of limit integrity due to operational mishap
$\bar{X}_t$: mechanical damage present at t
D_{5t} : damage size in region 5
B_{Lt} : limit integrity at t
$\bar{H}$: damage not detected
$\bar{X}_{OT}$: operational defect 1 at T
T_{1T} : type of defect is 1 at T
V_T : degree of defect at T.

If we instead consider the case with no mechanical damage, the equation becomes

$$P(\bar{U}_{LO}) = P(X_t \bar{X}_{OT} T_{1T} V_T \bar{B}_{LO})$$

(3.18)

Equation (3.17) describes the following basic events: mechanical damage is inflicted; it is not detected, and an operational mishap occurs. Equation (3.18) represents an operational violation in an otherwise 'flawless' structure, which is a relatively simple event. The next example starts with a numerical evaluation of Equation (3.18).

Example 3.10   Equation (3.18) can be expanded as

$$P(\bar{U}_{LO}) = P(\bar{B}_{LO} \mid X_t \bar{X}_{OT} T_{1T} V_T) \cdot P(V_T \mid T_{1T} \bar{X}_{OT}) \cdot P(T_{1T} \mid \bar{X}_{OT}) \cdot P(\bar{X}_{OT}) \cdot P(X_t)$$

It is assumed that Six-Sigma applies to the process for controlling load level to limit, and the effects of anomalous loads are determined by the discussion in Example 3.3. The first factor is assumed to be based on a 20% exceedance

$$P(\bar{U}_{LO}) = 0.0014 \cdot 0.4 \cdot 0.4 \cdot 3.4 \cdot 10^{-6} \cdot 0.1 = 0.8 \cdot 10^{-11}$$

which, for a 40% exceedance, becomes $1.7 \cdot 10^{-10}$. The total effect of operational defects will be assessed at the end of this section.

3.5.2 *Operational violations of the environmental envelope*

The structural design of composite structure is based on a thermal limit, T_{MAX}. It is important that this quantity is well defined in the requirements and design criteria. Very often, composite design data are based on the following integrity situation:

$$P(\bar{U}_U) = P(\bar{B}_U S_D S_E S_S) = P(\bar{B}_U \mid S_D S_E S_S) \cdot P(S_D) \cdot P(S_E) \cdot P(S_S) \qquad (3.19)$$

Here,

the first factor is loss of ultimate integrity with:

a state of damage, S_D

a state of environment, S_E

a state of a locally disturbed stress field, S_s, e.g. an open hole stress concentration; followed by the probabilities of the different states.

The state of damage to be associated with loss of ultimate integrity is based on the premise that if damage cannot be detected it must be accounted for in the design; this leads to the practice of accounting for damage in both strength and stability.

The environmental state requirements include a realistic definition of maximum design temperature. Recent investigations have revealed temperatures as high as 220°F in wing center sections after several hours of running the air conditioning. The delineation between action taken to ensure cooling off (avoidance), and accounting for the maximum temperature in the design, has to be firmly defined.

One may question consistency in today's design practices in combining stability and maximum temperature, especially as buckling occurs with B-value panel data, 'only' being adjusted for temperature effects, for no damage present and for uniform local stress/strain fields, especially as it has a much greater probability of occurring than, for example, 'open-hole, hot, wet' compression. We will illustrate this situation in the next example (Example 3.11).

Example 3.11   We assume that the temperature reduces the failure values for instability by 20%. Figure 3.5 illustrates the situation when B-values are used. We assume normally distributed data, and a design data comparison for ultimate loads.

The RT distribution has a mean, μ, and a B-value of 0.87μ, ($C_V = 0.1$), and the T_{MAX} dependent distribution is assumed to have a mean of 0.80μ. The normal distribution yields

$$\Phi\left(\frac{0.87\mu - 0.80\mu}{0.80\mu C_V}\right) = \Phi(0.875) = 1 - \Phi(-0.875) = 0.81$$

It is clear that the value of reduction can be serious and a threat to integrity.

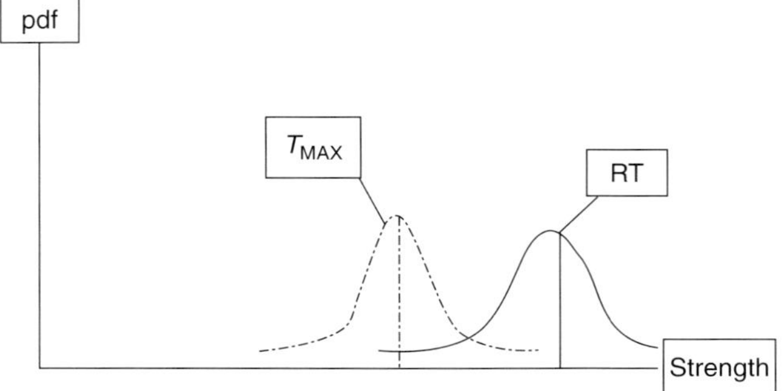

Figure 3.5 Comparative buckling at RT and maximum temperature.

The limit value results in

$$\Phi\left(\frac{\dfrac{0.87\mu}{1.5} - 0.80\mu}{0.80\mu C_V}\right) = \Phi(-2.75) = 0.003$$

So even in the 'legal' range we find that the probability of loss of limit integrity and failure is

$$P(\bar{U}_{LT}\bar{A}_T) = P(\bar{B}_{LT}\bar{C}_{TL}X_T X_{OT}) = P(\bar{C}_{LT}) \cdot P(\bar{B}_{LT} \mid X_T X_{OT}) \cdot P(X_T) \cdot P(X_{OT})$$
$$= 3.3 \cdot 10^{-5} \cdot 0.3 \cdot 10^{-2} \cdot 0.99 \cdot \sim 1 = 10^{-7}$$

This value certainly indicates an unacceptable reduction when the probability of exceeding limit load is $3.3 \cdot 10^{-5}$.

Another not infrequent practice is to use mean values as design data for buckling, the normal distribution yielding

$$\Phi\left(\frac{\mu - 0.80\mu}{0.80\mu C_V}\right) = \Phi\left(\frac{1.25 - 1.00}{0.1}\right) = 1 - 0.006 = 0.994$$

The limit value results in

$$\Phi\left(\frac{\dfrac{\mu}{1.5} - 0.80\mu}{0.80\mu C_V}\right) = \Phi(-1.7) = 0.04$$

The analogous value for loss of limit integrity and failure is

$$P(\bar{U}_{LT}\bar{A}_T) = 3.3\cdot10^{-5}\cdot0.4\cdot10^{-1}\cdot0.99 = 1.4\cdot10^{-6}$$

This numerical demonstration highlights the importance of dealing with the total picture on a case-by-case basis so that both the design criteria and operational boundaries can get the benefit of a coordinated solution.

3.5.3 Avoidance of environmental extremes

The remaining extremes mentioned in the introduction – extreme wind gusts (turbulence), wind shear, hailstorms and thunderstorms – all have to be dealt with through avoidance by predictions, reporting, information, and coverage that is reliable enough to keep the probability of total unsafe operation below its assigned maximum for total structural probability. This will be dealt with in the summary.

3.5.4 Ground damage – occurrence and reporting

Ground damage and the threats from it come in many forms, each deserving its own treatment. Severe ground damage causing loss of limit integrity, not initially detected or reported or inaccessible to 'walk-around' inspections, requires the most careful attention because either its occurrence must be made very improbable through very reliable quality assurance and process control, or the structure must be designed to be very damage resistant, if not fail-safe.

Equation (3.20) describes probabilities and events involved in damage undetected till T_1:

$$P(\bar{U}_{LT}\bar{H}_{T_1}) = P(\bar{B}_{LT}\bar{X}_{OT}T_1D_i\bar{H}_T\bar{R}_T A_{T_1}) \tag{3.20}$$

The events involved are:

$\bar{B}_{LT}$: the residual strength is less than limit load requirement, *LLR*, at *T*

$\bar{X}_{OT}$: a mishap in operation at *T*

T_1 : type of damage, 1 is debond or delamination

D_i : damage size region *i*

$\bar{H}_T$: damage not initially detected at *T*

$\bar{R}_T$: incident not reported at *T*

A_{T_1} : the structure survives till T_1.

Equation (3.20) can be expanded as

$$\begin{aligned}
P(\bar{U}_{LT}\bar{H}_{T_1}) = {}&P(\bar{B}_{LT} \mid \bar{X}_{OT}T_1D_i\bar{H}_T\bar{R}_T A_{T_1}) \cdot P(A_{T_1} \mid \bar{H}_T\bar{X}_{OT}T_1D_i\bar{R}_T) \\
&\cdot P(\bar{H}_T \mid \bar{X}_{OT}T_1D_i\bar{R}_T) \cdot P(\bar{R}_T \mid \bar{X}_{OT}T_1D_i) \cdot P(D_i \mid T_1\bar{X}_{OT}) \\
&\cdot P(T_1 \mid \bar{X}_{OT}) \cdot P(\bar{X}_{OT})
\end{aligned} \tag{3.21}$$

We will use the next example to explore what can be expected from the probability of an unsafe state due to this kind of joint event.

Example 3.12   The first factor on the right-hand side of Equation (3.21) is based on the use of residual strength B-values and therefore 10^{-1}. The second factor is based on:

$$\text{first flight after damage:} (\bar{p}_D + p_D\bar{p}_R)p_S + p_D p_R;$$

$$\text{consecutive flights:} [p_D + \bar{p}_D p_S]^{n-1}$$

where there are n flights between T and T_1. The following assumptions are used:

$$\text{Probability of detection } p_D = 0.95$$
$$\text{Probability of detection } p_R = 0.50$$
$$\text{Probability of detection } p_S = 0.99$$

and the factor becomes, for example for 30 flights:

$$[(0.05 + 0.95 \cdot 0.50) \, 0.99 + 0.95 \cdot 0.50] \, (0.95 + 0.05 \cdot 50)^{29}$$
$$= (0.525 \cdot 0.99 + 0.475) \cdot 0.48 = 0.48$$

and the probability of detection before 30 flights is $1 - 0.05^{30} \approx 1$.

The third factor, if based on $I = 5$, is: 10^{-3}. The fourth factor is pessimistically assumed to be 0.5.

The fifth factor is assumed to be 10^{-1}. The sixth factor is 0.5 and the last is P_G. The total becomes:

$$P(\bar{U}_{LT}\bar{H}_T) = 10^{-1} \cdot 0.48 \cdot 0.5 \cdot 0.5 \cdot 10^{-1} \cdot 0.5 \cdot P_G \approx 0.6 \cdot 10^{-3} \cdot P_G = 6P_G \cdot 10^{-4}$$

If the operational process has a Six-Sigma quality we get:

$$P(\bar{U}_{LT}\bar{H}_T) = 6 \cdot 3.4 \cdot 10^{-6} \cdot 10^{-4} = 20.4 \cdot 10^{-10} = 0.2 \cdot 10^{-8}$$

If we now instead look at the joint event that includes detection and reporting at T we get:

$$P(\bar{U}_{LT}H_T R_T) = P(\bar{B}_{LT} \mid \bar{X}_{OT}T_1 D_5 H_T R_T) \cdot P(H_T \mid \bar{X}_{OT}T_1 D_5) \cdot P(R_T \mid H_T \bar{X}_{OT}T_1 D_5)$$
$$\cdot P(D_5 \mid T_t \bar{X}_{OT}) \cdot P(T_1 \mid \bar{X}_{OT}) \cdot P(\bar{X}_{OT})$$

Which then analogously becomes

$$P(\bar{U}_{LT}H_T R_T) = 10^{-1} \cdot 1 \cdot 0.5 \cdot 10^{-1} \cdot 10^{-1} \cdot 3.4 \cdot 10^{-6} = 1.7 \cdot 10^{-9} \approx 0.2 \cdot 10^{-8}$$

This set of numerical examples illustrates the point that emerges often in the safety assessment of composite structure that by nature is case-by-case dependent, and which very often must be studied for many different scenarios.

3.6 REQUIREMENT TRANSGRESSIONS

Regulations, design requirements, criteria and design objectives, and the management of uncertainties are very important aspects of safety. The importance of their role in assuring safe structure is the main rationale for identifying all of them as parts of structural safety management.

3.6.1 Government regulations

The importance of enforced, up-to-date standards and regulations cannot be exaggerated. The role they play is an important one owing to the potential results of:

- missing regulations
- misleading regulations
- contradicting regulations
- 'drift' of requirements of regulations.

All of the above-listed items can have far-reaching consequences for structural safety. First, 'missing regulations' are very much the rule, not the exception, especially for transport-category composite aircraft. Very little has been achieved in establishing a baseline for composite structure beyond advisory circulars, which are not a minimum standard as is the case for Federal and International regulations.

Composite structure has, over time, involved the use of a large number of different materials that have different response behaviors, failure mechanisms and types of criticalities, (e.g. modes of failure), and there is a lack of an empirical database because of ever-continued usage of new materials and practices. This also applies to 'hybrid' structures, (combinations of composites and metals), which often behave dramatically differently from the individual components. Composites requine the application of specific regulations; the operational use of hybrid structures also needs to be regulated.

Secondly, 'misleading regulations' are a threat both to design criteria and design data. For example, the influence of maximum temperature on both design criteria and design data and its impact on safety is in need of regulations that control both the process for determining the maximum temperature and the probability of its being reached in service. The special considerations relating to hybrid structures should also be regulated from a safety standpoint.

Thirdly, looking at 'contradicting regulations', even here we have practices evolving in a non-consistent manner because of the lack of composites regulations; therefore *de facto* designs are based on old regulations that deal with metal which means that limit load concepts come in many versions. Presently, FAR 25 defines limit load as:

> The largest load expected in service.

Damage tolerance practices have established limit loads in two different versions:

> Limit load is defined as the largest external load
> encountered once in the life of an aircraft.

> Limit load is defined as the largest external load
> encountered once in the life of a fleet of aircraft.

From a safety standpoint the regulatory focus should be on the maximum, internal loads at the detail design points (DDPs) and occurring with a probability of 1 during the service life of the structure. Backman (2005) discusses this subject from a design standpoint; this section illustrates the general need for detailed definitions and the value of having a solid baseline for damage tolerance in critical composite structure. The next example discusses the detailed consequences of some of the options.

Example 3.13 We start with the definition of the largest order statistic at a point and consider n flights the basis for a sample, which gives us the following equation for the statistic:

$$g_n(m) = n[F(m)]^{n-1} \cdot f(m)$$

where $F(m)$ is the distribution function for the sample and $f(m)$ is the probability density function. We now ask the question: 'What is the probability that the maximum is larger than (LLR), limit load requirement? $\Pr\{LLR \leqslant L \leqslant \infty\}$

$$p = \int_L^\infty n[F(m)]^{n-1} f(m) \cdot dm = 1 - F^n(m)$$

The numerical evaluation gives the following results:

n	$F^n(L)$	p
1	0.999	10^{-3}
1000	0.905	0.095
30 000	~0	~1
1	0.99999	10^{-5}
1000	0.987	0.013
30 000	0.63	0.370

This table seems to indicate that this numerical range for limit in one flight would be too high.

If we instead look at the definition that says that a random value for exceeding limit is:

$$\Pr\{L \geq LLR\} = \frac{1}{30\,000} = 3.33\cdot10^{-5}\text{ for a random flight, then we have}$$

$$\Pr\{L \leq LLR\} \approx 1 - 3\cdot10^{-5} \Rightarrow$$
$$\Pr\{L \geq LLR \text{ in } 30\,000 \text{ flights}\} = (1 - 3\cdot10^{-5})^{30\,000} = 0.41.$$

Thus the probability that the maximum load will exceed limit in less than 30 000 flights is:

$$\Pr\{L \geq LLR \text{ in less than } 30\,000 \text{ flights}\} = 0.59$$

which ought to lead to the question: 'what should the total probability of "an unsafe state" be?

Equation (3.2) describes the contributions to loss of limit integrity; before we continue the total picture, we should investigate the design data part of the requirements.

A common practice is to use an extrapolation of the requirements in the 'metal world' leading to the use of B-values for the design data, provided that the structure involved is fail-safe. This, together with a philosophy to include 'very conservative design situations', has promoted the following formulation for design data:

The probability associated with ultimate strength is:

$$P(\bar{B}_U S_D S_E S_S) = P(\bar{B}_L \mid S_D S_E S_S) \cdot P(S_D) \cdot P(S_E) \cdot P(S_S) \tag{3.22}$$

$\bar{B}_U$: strength $\leq$ ultimate load requirement, $ULR = 1.5 \cdot LLR <$ for linear structure
S_D : state of damage
S_E : state of environment, e.g. maximum temperature
S_s : state of local stress.

The next example illustrates an often-used practice.

Example 3.14   For open-hole hot compression we would produce the following allowable values:

$$P(\bar{B}_U S_D S_E S_S) = 10^{-1} \cdot 10^{-2} \cdot 10^{-3} \cdot 10^{-1} = 10^{-7}$$

The first factor is the B-value definition
The second represents the state of damage included

The third is the probability of, for example, maximum temperature

The fourth is the probability that a mechanical fastener produces open-hole effects

This clearly represents both very conservative designs and nebulous requirements.

Regulations should be available to describe definitions and requirements.

3.6.2   Design criteria

Design criteria are a natural complement to regulations. They tend to introduce many interpretations of regulations, be supported by valuable service experience and present a practical design foundation for both traditional and new aircraft.

The application of old regulations to new aircraft requires a thorough adaptation that needs implementation of a disciplined process and thoughtful service monitoring to verify applicability.

To change from requirements validated by experience, or simply to ignore them (e.g. B-value design data quality, or fail-safe detail design requirements), demands an extensive validation to demonstrate compliance if new circumstances emerge.

The next section contains a summary of the effects and threats associated with the different elements of safety.

3.7   EFFECTS OF DIFFERENT ELEMENTS OF SAFETY

The effects created by different elements appear to belong to a few common categories, all of which have been investigated in Backman (2005) and above in this chapter.

3.7.1   Design and actions

The influence of damage dominates limit integrity-based design. The tools are:

- minimizing the probability of an unsafe state (account for mechanical damage);
- conducting effective walk around inspections wherever possible;
- designing fail-safe structure;
- maintaining high-quality, scheduled inspection methods.

All these measures involve, in the main, structural properties and damage and must be monitored and modified during service. The only other threat belonging to this arena is degradation due to, for example, elevated temperature which, in today's technological environment, must be dealt with through safe-life design.

3.7.2   Manufacturing mishaps and actions

There are two dominating categories of effects due to manufacturing mishaps, structural property reduction and spurious internal loads. Both can be either local, (only one or a few DDP's affected) or PSE-wide. The effects involved can either be uninterrupted to end of service, or can take place at discrete points in time. The approaches needed to protect the level of safety are:

- minimizing the probability of unsafe states through effective process control;
- conducting high-quality walk-around inspections whenever possible;
- designing fail-safe structures and providing an adequate, ultimate, margin of safety, which, considering that safe composite structure often is damage tolerance critical, does not produce any weight penalty;
- maintaining high-quality scheduled inspection methods.

These measures are very similar to those contained in the previous section, and illustrate the similarities between the design and manufacturing element requirements.

3.7.3 Maintenance defects and actions

The same two dominating categories apply to this element of safety, i.e. structural property reductions and spurious loads. However, defects in the detection process result in more missed damage and larger damage sizes, which must be incorporated into the service monitoring process to separate cases isolated from the ubiquitous.

3.7.4 Operational defects and actions

This element of safety mostly is influenced by short-term events. The most dominant categories are:

- pilot failure to avoid exceeding limit load; spurious loads;
- failure to avoid extreme environments; reduced structural properties;
- failure to detect or report ground damage (inc. improved walk-around quality); reduced strength.

The natural way to ensure and maintain safety in operation involves education, display technology, warning and reporting systems for reporting threats and accidents, and improved ground technology, for instance wind-shear-related detection requirements.

The part of the operational process set that applies to structural safety design includes reporting and creating an 'environment' supporting reporting of mishaps and collecting the database that can make regulations and design criteria more realistic.

The walk-around inspection quality improvement requirements should be developed. They should also focus on ground equipment that facilitates access to presently inaccessible parts of the structure.

This element, when well coordinated with engineering requirements, could produce more up-to-date regulations, better design criteria and increasingly safe performance.

3.7.5 *Requirements element*

The nature of this element is different from the others in terms of its effects and timing. However, its importance is huge. The thought of embarking on a process of designing, building and fielding new aircraft, with new materials, new processes and new requirements, while lacking pertinent service experience and having no regulations applicable is contrary to safety in innovation. This could easily totally negate the safety-driven actions, taken under the auspices of the other elements of safety. The processes associated with interpretations and updating of regulations and formulation design criteria must be of the same quality as the processes related to the other elements.

A few examples can be drawn on that illustrate this point. The requirements for composite and hybrid structures need: (1) special regulations and requirements for, for example, design data, damage tolerance data, fail-safety; and (2) definitions for, for example, limit load and maximum temperature and, finally, processes for defining damage size and detectibility.

The next chapter investigates measures and scenarios that should be considered in establishing safety levels for all PSEs.

3.8 CONCLUSIONS

This chapter is focused on a review of defects associated with different elements of safety. The review reveals defects, the effects of which are associated with the reaching of limit load. There are defects that are limited to a small area, perhaps as limited as a detail design point (DDP) (e.g. a defect of a fastener installation) and there are others that apply to a sub-assembly or a PSE (e.g. process failure of a wing skin), and which also apply to many DDPs. There are defects that stay 'active' through the life of the vehicle, e.g. spurious, internal loads that are built-in and change the frequency of reaching limit, internal load.

The importance of this is that the design requirements change from PSE to PSE and sometimes require a detailed understanding of the defects at hand, the safety requirements, the process failures and the extent of the defective structure.

The effective long-term objective for the Six-Sigma processes is defect-free below Six-Sigma and a gradual growth from 6σ to 7σ of 0 to 10% with a resulting process capability of $0.5 \cdot 10^{-12}$.

Chapter 4
Effects on the Probability of a Safe State

The design investigation in Backman (2005) concluded that safe states play an important role in the design of safe structure. This chapter complements the total picture with the effects from other elements, studying different scenarios. Backman (2005) identified requirements that dominated safety levels:

- low probability of an unsafe state, which influences the quality of design data (e.g. B-value residual strength), damage resistance, maximum growth rates, maximum damage size and inspection quality;
- high-quality 'walk-around' inspections;
- fail-safety (e.g. for structures with 'hard-to-detect' defects).

The definition of safe structure that was used in this context:

$$P(S_S) = P(S_D \ S_M \ S_I \ S_O \ S_R) \tag{4.1}$$

This equation states that the probability of a safe structure, S_S is the probability of the joint event:

$$
\begin{aligned}
S_D &\quad : \quad \text{safe design} \\
S_M &\quad : \quad \text{safe manufacturing} \\
S_I &\quad : \quad \text{safe maintenance} \\
S_O &\quad : \quad \text{safe operation} \\
S_R &\quad : \quad \text{safe requirements}
\end{aligned}
$$

Equation (4.1) can be expanded as:

$$P(S_S) = P(S_D \mid S_M S_I S_O S_R) \cdot P(S_M \mid S_I S_O S_R) \cdot P(S_I \mid S_O S_R) \cdot P(S_O \mid S_R) \cdot P(S_R)$$

and the probability of an 'unsafe state' is:

$$P(\bar{S}_S) \approx P(\bar{S}_D \mid S_M S_I S_O S_R) + P(\bar{S}_M \mid S_I S_O S_R) + P(\bar{S}_I \mid S_O S_R) + P(\bar{S}_O \mid S_R) + P(\bar{S}_R) \tag{4.2}$$

This equation describes the different contributions from the elements, and their relative contributions will be investigated in the following sections.

This chapter will continue to investigate different process violations, with emphasis on the critical ones and what could be considered the maximum tolerable value of defects. The focus will be on structural effects. Here it is important to recognize that the term dealing with the probability of unsafe requirements in Equation (4.2) can be expressed as:

$$P(\overline{S}_R) = P(\overline{R}_P) + P(\overline{A}_P)\tag{4.3}$$

Here, two events are included:

$\overline{R}_P$: unsafe process requirements
$\overline{A}_P$: unsafe process applications.

This requires appropriate regulations, criteria and objective discussions, which in turn requires thorough insights into all the aspects of structural design and process development and applications.

4.1 MANUFACTURING AND PROCESSING DISCREPANCIES

Detection by observation and inspection is a very desirable design objective. However, types of process mishaps exist, the results of which are difficult to detect using current methods. It is very important to identify them and classify the threats involved. In order to study different scenarios we will now illustrate the probability of failure for a few defects assuming normal distributions, where k_x is property reduction and k_y load increase:

$$\mu_g = k_x\mu_x - k_y\mu_y \Rightarrow k_x 1.13LLR - k_y 0.7LLR \quad \text{and}$$

$$\sigma_g = C_V \cdot LLR[k_x^2 \cdot 1.13^2 + k_y^2 \cdot 0.49]^{1/2} \quad \text{and} \quad \Phi\left(\frac{-\mu_g}{\sigma_g}\right) = \Phi(t)$$

k_x	k_y	μ_g	σ_g	t	$\Phi(t)$
1.0	1.0	0.43	0.134	−3.20	0.00069
0.9	1.0	0.317	0.123	−2.58	0.00460
0.8	1.0	0.204	0.090	−2.26	0.01200
1.0	1.1	0.36	0.136	−2.64	0.00450
1.0	1.2	0.29	0.141	−2.05	0.02000

The table, which shows the probability of failure results for different defect values, is used in the numerical evaluation below.

4.1.1 *Property reduction*

Reduction in structural properties due to failure of process, e.g. for loss of strength (and equivalent), can be expressed as the probability of an unsafe state:

$$P(\bar{S}) = P(\bar{X}_M T_1 V_1 X \bar{H}_0 \bar{B}_L) = P(\bar{B}_L \mid X\bar{X}_M T_1 V_1 \bar{H}_0) \cdot P(\bar{H}_0 \mid \bar{X}_M T_1 V_1)$$
$$\cdot P(V_1 \mid \bar{X}_M T_1) \cdot P(T_1 \mid \bar{X}_M) \cdot P(\bar{X}_M) \qquad (4.4)$$

Here the following events are involved:

$\bar{X}_M$: manufacturing mishap
T_1 : type of mishap, e.g. reduction in strength
V_1 : size of reduction
X : no mechanical damage
$\bar{H}_0$: inspection does not detect anything
$\bar{B}_L$: limit integrity lost.

The next example illustrates the problem of non-detection of mishaps.

Example 4.1 We assume that the strength is reduced by 20%, and that the process has a Six-Sigma quality. From a previous example we know that the probability of failure is 0.0052 if we assume normal variables. Equation (4.4) yields for the probability of an unsafe state:

$$P(\bar{S}) = 0.82 \cdot 0.99 \cdot 0.2 \cdot 0.6 \cdot 3.4 \cdot 10^{-6} = 0.33 \cdot 10^{-6}$$

and the probability of failure in n flights p_{fn} is

n	p_s	p_{fn}
1	0.9946	0.0052
100	0.5800	0.4200
1000	0.0081	0.9919
3000	$\sim 10^{-8}$	~ 1

The table, showing the probability of failure, indicates that a fail-safe design must be used to deal with this type of hard-to-detect defect (20% reduction in strength).

A prudent choice of critical limit damage sizes could lead to an ultimate margin of safety greater than 20%, which means that the ultimate capability of the 'remaining' structure would have the potential of inherent fail-safety, but load-path failure in compression (due to limit loads), especially in composites structure, is a very difficult subject owing to the failure dynamics involved and will potentially require design development testing, design data testing, and compliance testing. The numerical range for 20% reduction and the associated probability of an unsafe state is described in the following equations:

$$\frac{F_L}{f_L} - 1 = 0 \Rightarrow \frac{F_U 1.5}{f_U \sqrt{r}} - 1 = 0$$

where r is the ratio between limit damage size and ultimate damage size if $r \geq 3.25 \Rightarrow MS \geq +20\%$

$$
\begin{aligned}
P(\bar{S}) = P(\bar{X}_M T_1 V_1 \bar{B}_L \bar{Y}_t R_t B_t \bar{H}_T) &= P(\bar{B}_L \mid \bar{X}_M T_1 V_1) \cdot P(V_1 \mid \bar{X}_M T_1) \cdot P(T_1 \mid \bar{X}_M) \\
&\quad \cdot P(\bar{X}_M) \cdot P(\bar{Y}_t \mid \bar{B}_L) \cdot P(R_t \mid \bar{Y}_t) \cdot P(\bar{H}_T \mid \bar{Y}_t R_t) \\
&= 0.82 \cdot 0.2 \cdot 0.6 \cdot 3.4 \cdot 10^{-6} \cdot 0.012 \cdot 0.5 \cdot 10^{-3} \\
&= 0.2 \cdot 10^{-11}
\end{aligned}
$$

This value at least supports a reasonable criterion for safety, and it illustrates how damage tolerance critical structure directly supports a fail-safe approach to structural protection from process failure. Reduced stiffness properties are, from a buckling standpoint, analogous to reduced strength properties. Damage tolerance (residual strength) and damage resistance, if reduced, and damage growth rates, if increased, can have a significant and detrimental influence on safety. The following examples will deal with the three latter properties.

Example 4.2   This example illustrates the results of a 20% reduction in residual strength. The probability of an unsafe state is:

$$
\begin{aligned}
P(\bar{S}) = P(\bar{X}_M T_{12} V_{12} \bar{X}_T D_5 \bar{H}_T \bar{B}_{LT}) &= P(\bar{B}_{LT} \mid \bar{X}_T D_5 \bar{X}_M T_{12} V_{12} \bar{H}_T) \\
&\quad \cdot P(\bar{H}_T \mid \bar{X}_T D_5) \cdot P(D_5 \mid \bar{X}_T) \cdot P(\bar{X}_T) \cdot P(V_{12} \mid \bar{X}_M T_{12}) \\
&\quad \cdot P(T_{12} \mid \bar{X}_M) \cdot P(\bar{X}_M)
\end{aligned}
\tag{4.5}
$$

A numerical evaluation assuming normal distributions yields:

$$P(\bar{S}) = 0.79 \cdot 10^{-3} \cdot 10^{-3} \cdot 10^{-2} \cdot 0.2 \cdot 0.2 \cdot 3.4 \cdot 10^{-6} \approx 10^{-15}$$

The resulting value is significantly less critical than the result in the previous example.

Example 4.3   We assume that the reduction in damage resistance is 20%, (enough to skip over the buffer zone for damage sizes between D_4 and D_5), which makes fail-safety a required design objective. It is assumed that load redistribution is possible when the pertinent load path fails. The guiding equation is:

$$P(\bar{S}) = P(\bar{X}_M T_{13} V_{13} \bar{X}_t D_{5t} \bar{B}_{Lt} \bar{Y} R \bar{H}_T) = P(\bar{B}_{Lt} \mid \bar{X}_t D_{5t} \bar{X}_M T_{13} V_{13}) \cdot P(D_{5t} \mid \bar{X}_t) \cdot P(\bar{X}_t)$$
$$\cdot P(V_{13} \mid T_{13} \bar{X}_M) \cdot P(T_{13} \mid \bar{X}_M) \cdot P(\bar{X}_M)$$
$$\cdot P(\bar{H}_T \mid \bar{Y}R) \cdot P(R \mid \bar{Y}) \cdot P(\bar{Y})$$
$$= 0.10 \cdot 10^{-2} \cdot 10^{-2} \cdot 0.2 \cdot 0.1 \cdot 3.4 \cdot 10^{-6}$$
$$\cdot 10^{-3} \cdot 10^{-1} \cdot 0.012$$
$$= 0.8 \cdot 10^{-17}$$

The critical part of the threat management used here is to have a capability designed into the structure that assures load redistribution when the critical load path fails under load (especially in compression). The only alternative is walk-around inspection to find the external damage before failure occurs. It seems that this solution to damage resistance has to come in and—with a test demonstration of the maximum reduction value allowable in manufacturing.

Example 4.4   The increase in damage growth rates will have to be limited to less than 1.5, provided both walk-around inspections and fail-safe design are applied. Practically, that would mean that damage size grows to maximum size in two intervals and the probability of an unsafe state would be

$$P(\bar{S}_3) = P(\bar{H}_2 \bar{U}_3 \bar{H}_3) = P(\bar{H}_2 \mid D_5) \cdot P(\bar{B}_L \bar{X}_t D_5) \cdot P(\bar{H}_3 \mid D_5)$$
$$= 10^{-3} \cdot 10^{-6} \cdot 10^{-3} = 10^{-12}$$

Here the size region D_5 is reached at T_2 (see Figure 4.1).

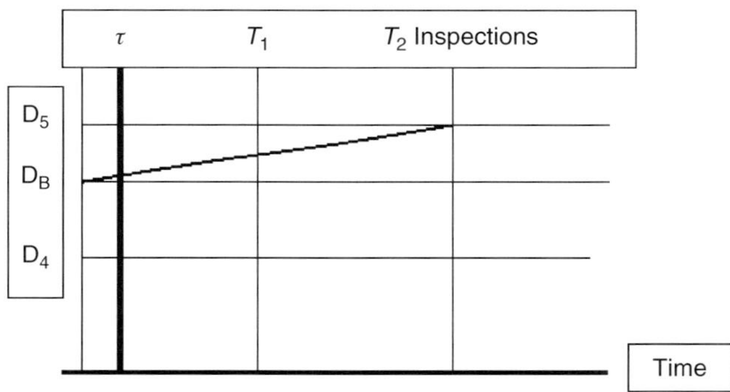

Figure 4.1 Increased damage growth rates.

The easiest protection against increased damage growth seems to be an efficient walk-around inspection as large, easily accessible damage is involved. Even if the probability of detection were only 0.5, the probability of being detected in less than or equal to n flights, p_{dn}, is:

n	$(1-0.5)^n$	p_{dn}
1	0.5	0.5
10	10^{-3}	0.999
20	10^{-6}	~1
30	10^{-9}	1

Therefore, walk-around inspection as an alternative to fail-safe design with loss of load path under limit load seems to be an effective choice.

4.2 MECHANICAL FLAWS AND DAMAGE

The preservation of limit load capability is a very important design objective for composite and hybrid structures. Refinements in damage tolerance of composite materials and structures have become a large part of ongoing innovation. Needs for safety-based design criteria with requirements for maximum damage sizes and probabilities of detection have become imperative, as has the need to categorize damage types and origins.

4.2.1 Location of defects

The location of damage is important, especially from the standpoint of whether it is accessible to walk-around inspection or not. This type of inspection is a highly effective tool in preserving safety by detection, as illustrated in Example 4.4. If such inspection is not possible, major, scheduled inspections must be resorted to.

Detection of damage of composite structure during in-service scheduled inspection depends on a combination of external and internal damage sizes. Many types of process failures occur in manufacturing, maintenance and operation, (e.g. de-laminations and de-bonds) that do not produce external damage but cause substantial internal destruction. These situations must be rendered very improbable through process control and quality control.

Figure 4.2 shows an example describing the influence of external damage, D_{Se}, and internal damage, D_{Si}, which can be used to determine regions where worst-case detection scenarios can be used as part of the definition of design criteria. A case-to-case dependent rigor is likely to be the realistic way to deal with very small external damage sizes. The alternative of fail-safe structure, which is a regulatory requirement if B-values are used, becomes a final safety bulwark. The conclusion is that fail-safe design and testing requirements must be a fundamental part of regulations, including rules for detailed compliance demonstrations.

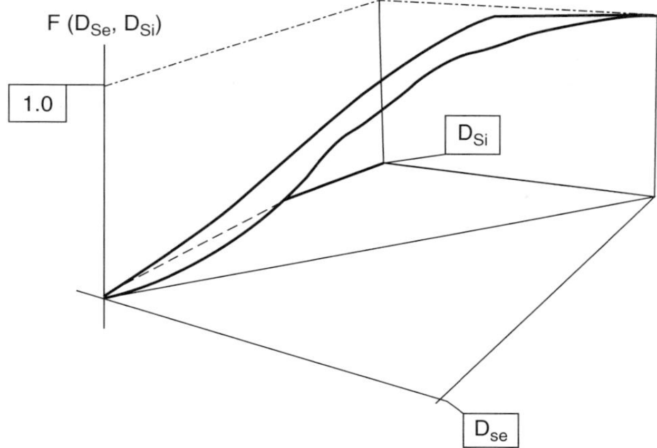

Figure 4.2 Detectibility distribution.

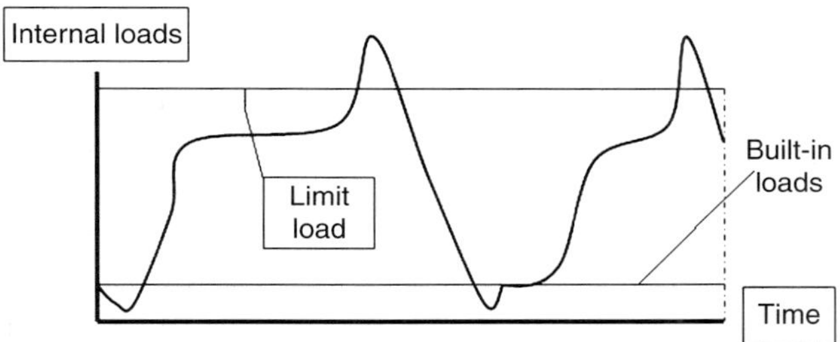

Figure 4.3 Internal loads and spurious loads.

4.3 'ADDED' SPURIOUS INTERNAL LOADS

'Built-in' internal loads (stresses) can be a result of defects in manufacturing processes (mismatches, clamp-up, faulty installation, etc.) or poor repairs (in relation to fastener installations, hole tolerances, geometry, etc.), in maintenance. There are two basic effects, built-in self-equilibrating internal loads that are constantly present, and erratic, internal design load distributions that create local exceedances. Both cause increases in maximum local operating loads.

Figure 4.3 illustrates the effects of 'built-in' loads – always present except for relaxation – in composite structure. The threat is constantly present and underscores the need for a 'crisp' definition of limit load.

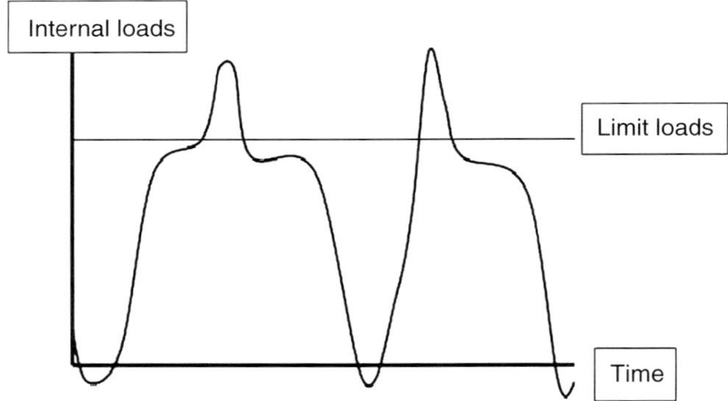

Figure 4.4 Limit exceedances due to defects.

Example 4.5 If the built-in internal loads correspond to the response to 20% of limit external loads, we find that

$$Pr\{0.80 \cdot LL < L_E < LL\} = Pr\{LL_I < L_I < 1.2LL_I\}$$

This equation shows that for a 20% built-in internal load to have the same probability of exceeding internal limit load by 20% of the external limit load it has to be in the range $(0.8 \cdot LL < L_E < LL)$.

Internal load maxima both for defects in manufacturing and for mishaps in operation are described in Figure 4.4.

The peaks are of two kinds. The one caused by mishaps in operation can result in internal loads exceeding limit. That caused by defects in manufacturing results in spurious internal load concentrations; the maximum shown with the occurrence of external limit loads depends on the definition of limit external loads.

The first three sections of this chapter cover the main effects of process defects, the next section establishing the basic threats to safety and the design features and process quality required.

4.4 SUMMARY OF EFFECTS

The previous three sections have identified the following fundamental effects of defects:

- degraded structural properties;
- excessive (above limit) internal loads;
- mechanical damage.

An assortment of examples has been used to illustrate bounds and suggested actions.

The chosen and recommended actions are:

- minimization of the probability of an unsafe state;
- fail-safe design;
- walk-around inspections, whenever possible;
- service monitoring and data analysis for risk management (updating and uncertainty reduction).

The remaining types of defect do not add any other approach to the overall preservation of structural integrity over and above the stringent process quality control of all applications. It becomes very important in material characterization and selection to realize that the process quality required in manufacturing has to be determined in the developmental phase and that the difficulty in achieving and maintaining this and the associated compliance and monitoring (to avoid and correct drift) constitute very significant structural safety issues.

4.5   INSTALLATION AND ASSEMBLY PROBLEMS

Problematic defects for both of these processes are:

- reduction of structural properties owing to compromised fastener properties, hole-quality, geometry, bond-line integrity and fastener selection;
- built-in internal loads due to mistakes in geometry and clamp-up;
- non-uniform internal load distributions due to incorrect hole-tolerances, varying bond-line thickness, variations in clamp-up and inadequate grip-length.

In summary, we find that defects in design data, induced internal loads and errors in internal loads distributions are all limited to 20% deviations of Six-Sigma process quality.

4.6   MAINTENANCE DEFECTS

Scheduled maintenance, inspection and repairs can lead to:

- exceedances of limit loads due to built-up resistance from inadequate scheduled maintenance or defects in repairs;
- reduction of structural properties due to incorrectly installed fasteners, faulty bond-lines or damage inflicted during the repair procedure;
- large-scale damage missed in inspections and threatening structural integrity.

The resulting defects include: internal load exceedances, inadequate structural property values and greater frequency of large damage size being present.

4.7 OPERATIONAL MISHAPS

The serious defects caused by external limit loads exceedances, failure to avoid environments outside of the design envelope and not detecting or reporting ground or flight damage can cause:

- larger-than-limit internal loads;
- structural properties below required design values;
- large damage sizes that are neither detected nor reported and remain a threat to the structural safety of a vehicle.

These discrete events can cause larger-than-limit internal loads, structural properties less than design values and structural damage sizes larger than the maximum for damage-tolerant structures.

4.8 REQUIREMENT FORMULATION

There are several sources of threats to safety on the requirement side of both composite and hybrid structures. Federal and International regulations covering these types of structure are very sparse. The practice of applying existing 'metal-relevant' rules, especially to composite structures, is often inadequate because of the differences in behavior between metallic and composite materials and structures. In addition, composites exhibit varied characteristics for different composites.

Definitions in FAR 25 of limit loads, environmental effects to consider in design, design values, damage resistance, damage growth rates, damage tolerance and complementing fail-safe criteria, etc., all need to be updated to reflect the true design challenges for composite and hybrid structures.

Many of the present practices that have evolved for composites are nebulous and often misleading. For example, limit loads must include a definition for internal loads and the current 'limit load is the largest load expected in service', must have a probabilistic complement to the design criteria, particularly for innovative, emerging designs.

The future must include safety-based structural design data, internal design loads and a robust definition of limit load. Damage resistance and fail-safe requirements for structures that fail or are damaged up to limit load must be added to complement design criteria, especially for damage-tolerant critical structure. Rules to determine damage growth rates for different situations are essential, as are rigorous definitions of maximum temperatures and moisture content.

The practices for composites-allowable values are reasonably consistent and could be extended to design data. They are based on the idea that if the internal ultimate loads exceed allowable ultimate values, then ultimate integrity, $\bar{U}_U$, is lost and, consequently, fail-safe integrity also. The following equation expresses the situation:

$$P(\bar{U}_U) = P(\bar{B}_U S_D S_E \bar{S}_S) = P(\bar{B}_U \mid S_D S_E S_S) \cdot P(S_D) \cdot P(S_E) \cdot P(S_S) \qquad (4.6)$$

Here the following events are included:

$\bar{B}_U$: Ultimate maximum, internal load is larger than the design value
S_D : State of damage
S_E : Environmental state
S_S : State of 'local' stress or strain concentration.

An illustration of this situation is included in the next example.

Example 4.6 We assume that the first factor in Equation 4.6 is required to be of B-value quality. The second one is 10^{-3}, and the third factor, representing maximum temperature, is 10^{-3}. Finally, the fourth represents a local internal load situation, analogous to an open hole. The situation would then describe 'hot, open-hole compression', and results in the following probability of losing ultimate integrity:

$$P(\bar{U}_U) = 10^{-1} \cdot 10^{-3} \cdot 10^{-3} \cdot 10^{-2} = 10^{-9}$$

This value seems to describe a very conservative approach to design data.

A prevailing practice involving buckling would present a very different level of criticality:

$$P(\bar{U}_U) = 0.5 \cdot 0.99 \cdot 0.99 \cdot 0.99 = 0.485$$

This result appears very unrealistic, mainly because it deals with average test results, no damage, no temperature and uniform internal loads. It appears that potentially critical design data would be a great deal more homogeneous should realistic regulations be in place.

Damage tolerance, commonly a design requirement for composites, presents a very similar situation. The probability of loss of limit integrity is described in Equation (4.7).

We assume that the region of damage sizes that represents damage tolerance requirements is D_5, the equation for loss of limit integrity is:

$$P(\bar{U}_L) = P(\bar{B}_L S_D S_E S_S) = P(\bar{B}_L \mid S_D S_E S_S) \cdot P(S_D) \cdot P(S_E) \cdot P(S_S) \qquad (4.7)$$

Here the first factor on the right-hand side is the allowable residual load value, S_D; $\overline{XD}_5$ is the state of damage; S_E is temperature; and S_S could represent 'severity'.

We will investigate a realistic set of numerical values for comparison in the following example.

Example 4.7 The investigation of safety levels reported in Backman (2005) indicates that a reasonable value for the first factor on the right hand side of Equation 4.7 is 10^{-1}; the second factor would be $10^{-2} \cdot 10^{-3}$; the third would represent a typical flight temperature, and a realistic estimate would be ~ 0.9; and if the fourth factor describes 'severity' the estimate would be of the order of magnitude of ~ 1.2. The resulting probability of loss of limit integrity is

$$P(\overline{U}_L) = 10^{-1} \cdot 10^{-5} \cdot 0.9 \cdot 1.2 \approx 10^{-6}$$

This result seems to support the assertion that safety requirements often result in

$$P(\overline{U}_L) \leq 10^{-6}$$

and B-value residual strength design values would be preferred, if limit damage size, D_5, is used as suggested in Backman (2005).

To illustrate this result we will assume normally distributed ultimate strength and design B-value data.

The B-value is calculated by

$$\Phi\left(\frac{B - \mu}{\sigma}\right) = 0.1 \Rightarrow B = 0.87\mu \quad \text{with } C_V = 0.1$$

and the limit probability becomes

$$\Phi\left(\frac{\dfrac{0.87\mu}{1.5} - \mu}{\sigma}\right) \Rightarrow t = -4.2 \Rightarrow \Phi(-4.2) = 10^{-5}$$

This limit probability derived from ultimate criticality seems to support, at least, the B-value residual strength quality.

The determination of a balanced value for the probability of loss limit integrity thus becomes the result of a balance between residual strength allowable value 'quality', limit damage sizes and the probability of inflicted damage in the damage-tolerance size region.

The entire process of determining the probability of loss of both ultimate and limit integrities should be the subject of composite structures regulations.

The definition of limit load should be complemented with a definition of internal loads limit, in order to create a safety-based design tool for damage-tolerant composite structure. The present definition in FAR 25 is based on external loads:

The largest load expected in service.

The design process uses the largest internal load at the detail design point (DDP) being considered. A vehicle with a service objective of n flights should for example have a probability of $1/n$ of reaching limit internal load during a random flight. That would result in a probability of 1 of reaching limit internal load once during the vehicle design objective period. This definition would allow a realistic process for development in safe design criteria.

The most important areas in producing and maintaining safe composite vehicle structures are limit and ultimate design values and design loads. They are also, together with damage, the important and realistic entities to monitor, analyze and update during mitigation of risk and uncertainty in service.

Present practices in composite design, both in producing design values and design loads, vary in many ways. The state of affairs in composites is such that more rigorous design data processes must be part of the requirements of the regulations. The definition of limit loads should be updated, and the use of regulations developed for metallic structures as a 'crutch' should be abandoned.

The defects in the process that involves the development and use of regulations are such that Six-Sigma is not within reach. A very dramatic example comes from the use of mean values for structural residual strength, which results in a limit integrity that is only met 50% of the time for symmetric data.

This situation makes the 'element of requirements formulation' the most important element of safety.

4.9 MONITORING, INTERPRETATION, REPORTING AND UPDATING

Innovation comes with a lack of service experience and consequently no empirical insights and no design databases. The 'composite world' exhibits a steady stream of new materials, new processes and new structural concepts and faces much uncertainty and many forms of risk, so the quest for structural safety starts with design decisions and continues until the end of the design objective.

The monitoring of airplanes consists very much in the detection and reporting of damage and interpretation of sizes and frequencies. The purpose of these activities is to compare reported damage sizes and detection against the design criteria that include the probability of occurrences and detection. Similar attention should be paid to reported spurious vibrations, which could be indications of local loss of stiffness. The use of monitoring in safety management must, among other things, be a search for and interpretation of damage sizes to assess risk and mitigate uncertainty, including finding failed load-paths in fail-safe structure.

Discovering and updating uncertainty in data, models, the nature of phenomena and requirements is a fundamental aspect of risk management and uncertainty reduction. It provides an opportunity to protect levels of safety, but also makes a great contribution to education and the long-term value of composites and other 'new' structures.

4.10 CONCLUSIONS

The following actions support the design of a safe structure:

- minimizing the probability of unsafe states (undetected loss of integrity);
- designing fail-safe structures;
- conducting walk-around inspections wherever possible;
- establishing the long-term Objective for Process Development;
- executing the following processes in service:

  Risk management driven by data-based probability of an unsafe state to restore an acceptable level of safety by adjusting the inspection approaches and periods; using the same data to reduce uncertainty whenever possible.

Chapter 5
Process Defects Affecting Quality of Structures

The processes constituting a threat to structural safety are a consequence of detail design (forced process choices) and of innovation producing an ever-increasing number of processes (alternative choices), which results in the need to evaluate risk, uncertainty and iterate design decisions. The result is that the design processes and 'optimizations' become increasingly complex and involve many different versions of safety management.

Safety management therefore becomes a philosophy of combining different elements, process qualities and process limits. It is a combined set of processes developed to enable the handling of any situation that arises. The elements of safety all need procedures to establish their own process quality, guard performance and implement continued improvements.

The 'total picture' also includes structural integrity, the primary considerations being:

- reduction of structural properties
- internal overloads
- mechanical damage.

The effects of all elements of safety can be found under these three headings. They can combine with moderate probabilities, or selected elements can combine with event probabilities close to the global requirements.

The probability of a specific defect can be analyzed using the example quality associated with the Six-Sigma quality; the probability of occurrence is

$$P(V_{1i}T_1\bar{X}_M) = P(V_{1i} \mid T_1\bar{X}_M) \cdot P(T_1 \mid \bar{X}_M) \cdot P(\bar{X}_M) \tag{5.1}$$

V_{1i} : value i of type 1 defect
T_1 : type 1 defect
$\bar{X}_M$: e.g. manufacturing defect present.

The first factor on the right-hand side of Equation (5.1) is a representation of three different size ranges, the smallest of which preserves structural integrity; the moderate range preserves fail-safe integrity; and a further very unlikely range either is not a factor or is handled through fail-safety or discrete source damage.

Mechanical damage is a very special case because it is part of the structural design criteria. The structural design, especially for composites, must include requirements for damage sizes, damage resistance, damage growth, damage tolerance and damage detection. Mechanical damage comes from many sources and can be expressed in terms of characteristics, location, size and detectibility. The event 'mechanical damage is present', $\bar{X}_D$, can be expressed as

$$
\begin{aligned}
P(\bar{X}_D) &= P(\bar{X}_D\bar{X}_M) + P(\bar{X}_D\bar{X}_I) + P(\bar{X}_D\bar{X}_O) + P(\bar{X}_D\bar{X}_R) \\
&= P(\bar{X}_D \mid \bar{X}_M) \cdot P(\bar{X}_M) + P(\bar{X}_D \mid \bar{X}_I) \cdot P(\bar{X}_I) \\
&\quad + P(\bar{X}_D \mid \bar{X}_O) \cdot P(\bar{X}_O) + P(\bar{X}_D \mid \bar{X}_R) \cdot P(\bar{X}_R) \\
&\quad + P(\bar{X}_D \mid \bar{X}_{DP}) \cdot P(\bar{X}_{DP})
\end{aligned}
\tag{5.2}
$$

The first term of the right-hand side of Equation (5.2) represents manufacturing-induced mechanical damage. This defect could be caused by tool-drops, local resin starvation, rips in bags, transportation accidents, inclusions and machine gouges. These types of defect are stochastically independent over the principal structural element (PSE). So, with 'n' detail design points (DDP's), the requirement at a DDP it is $1/n$ of what the PSE requires. The probability of an unsafe state is

$$
\begin{aligned}
P(\bar{X}_{Dt}\bar{X}_M H_T \bar{R}_{PT}\bar{U}_T \bar{X}_{IT}V_i) &= P(H_T \mid \bar{X}_{Dt}) \cdot P(\bar{X}_{Dt} \mid \bar{X}_M) \cdot P(\bar{X}_M) \\
&\quad \cdot P(\bar{U}_T \mid \bar{R}_{PT}\bar{X}_{IT}V_i) \cdot P(V_i \mid \bar{R}_{PT}\bar{X}_{IT}) \\
&\quad \cdot P(\bar{R}_{PT} \mid \bar{X}_{IT}) \cdot P(\bar{X}_{IT})
\end{aligned}
\tag{5.3}
$$

$$
\begin{array}{lll}
\bar{X}_{Dt} &:& \text{mechanical damage present at } t \\
\bar{X}_M &:& \text{manufacturing defects are present} \\
H_T &:& \text{mechanical damage (in D}_4\text{) detected at } T \\
\bar{R}_{PT} &:& \text{repair process at } T \text{ is defective (type of maintenance process)} \\
\bar{U}_T &:& \text{structural integrity lost at } T \\
\bar{X}_{IT} &:& \text{faulty maintenance process} \\
V_i &:& \text{value of defect has reached failure level.}
\end{array}
$$

The numerical possibilities are explored in the next example.

Example 5.1 Equation (5.3) is evaluated assuming the largest-scale 'safe' damage (in D_4) is present. The numerical situation is compatible with the design and process requirements objective used in Backman (2005), and the PSE is assumed to possess 20 DDPs:

$$
P(\bar{S}_T) = 0.5 \cdot 0.5 \cdot 10^{-1} \cdot 10^{-2} \cdot 10^{-3} \cdot 0.5 \cdot 10^{-6} \cdot 20 = 2.5 \cdot 10^{-12}
$$

It is important to accept that this example represents a case-to-case situation that, if not detected, constitutes loss of fail-safe integrity.

The second term describes the effects of maintenance defects (e.g. damage inflicted in damage removal, de-bonding, de-lamination due to substandard fastener installations, etc). The consequence can be loss of fail-safe integrity. The following equation describes the effect of adjacent failure:

$$P(\bar{H}_L \bar{Y}_{At} \bar{A}_A \bar{A} \bar{U}_{FS} \bar{X}_I V_i) = P(\bar{H}_L \mid \bar{U}_{FS} \bar{X}_I) \cdot P(\bar{U}_{FS} \mid \bar{X}_I V_i) \cdot P(V_i \mid \bar{X}_I) \cdot P(\bar{X}_I$$
$$\cdot P(\bar{A} \mid \bar{A}_A \bar{Y}_{At} \bar{U}_{FS} \bar{X}_I) \cdot P(\bar{A}_{At} \mid \bar{Y}_{At} \bar{U}_{FS} \bar{X}_I) \cdot (\bar{Y}_{At}) \qquad (5.4)$$

$\bar{H}_L$	:	defects in repaired load path was not detected
$\bar{Y}_{At}$	:	damage inflicted in adjacent load path at t
$\bar{A}_A$	:	adjacent load path failed
$\bar{U}_{FS}$	:	repaired load path does not have fail-safe integrity
$\bar{X}_I$	:	process fault in maintenance.

Example 5.2 This example illustrates how mechanical defects in the processes of the elements of safety contribute to loss of fail-safe integrity. Equation (5.4) is used

$$P(\bar{S}_t) = 10^{-2} \cdot 0.39 \cdot 10^{-2} \cdot 10^{-6} \cdot 1 \cdot 0.2 \cdot 10^{-2} = 0.8 \cdot 10^{-13}$$

The main consequence, then, is that an unsafe state has been reached and with a proposed probability of reaching a limit of $0.33 \cdot 10^{-4}$ the probability of survival for 3000 to 6000 flights is 90–80%; the safety has to be assured by walk-around inspections.

5.1 DEFECTS OF MANUFACTURING

The processes of the elements of safety that can produce a reduction in structural properties are:

- manufacturing processes
- maintenance processes
- operation
- requirements.

5.1.1 Structural properties affected by manufacturing

The structural properties involved are:

- strength, S_R
- stiffness, E

- residual strength, R_S
- damage resistance, D_R
- damage growth, D_G
- damage accumulation, D_A.

The probability of safe structural properties, S_P is:

$$P(S_P) = P(S_R ER_S D_R D_G D_A) = P(D_A \mid D_G D_R R_S ES_R) \cdot P(D_G \mid D_R R_S ES_R)$$
$$\cdot P(D_R \mid R_S ES_R) \cdot P(R_S \mid ES_R) \cdot P(S_R \mid E) \cdot P(E)$$

$$(5.5)$$

and the probability of an unsafe state is

$$P(\overline{S}_P \mid \overline{X}_M) \approx [P(\overline{D}_A \mid D_G D_R R_S ES_R) + P(\overline{D}_G \mid D_R R_S ES_R) + (\overline{D}_R \mid R_S ES_R)$$
$$+ P(\overline{R}_S \mid ES_R) + P(\overline{S}_R \mid E) + P(\overline{E})] \cdot P(\overline{X}_M)$$

$$(5.6)$$

This equation yields an example that can be evaluated on a case-by-case basis.

It would seem that the following results are common:

$$P(\overline{S}_P \mid \overline{X}_M) = 0 + 0 + 0 + 2 \cdot 10^{-12} + 3 \cdot 10^{-14} + 10^{-12} \approx 3 \cdot 10^{-12}$$

The next example (Example 5.3) illustrates a way to deal with spurious effects for structural properties due to manufacturing defects.

Example 5.3 This example uses Equation (5.6) to illustrate orders of magnitude. The first term on the right-hand side expresses the probability of anomalous damage-accumulation with safe damage growth, safe damage resistance, safe residual strength, safe stiffness and safe strength. This describes a very unlikely, combined event caused by manufacturing defects. The same argument is true for terms two and three.

If we now narrow Equation (5.6) to manufacturing defects only, we get

$$P(\overline{S}_P \overline{X}_M) = P(\overline{S}_P \mid \overline{X}_M) \cdot P(\overline{X}_M) = 3 \cdot 10^{-12} \cdot 3.4 \cdot 10^{-6} \approx 10^{-17}$$

Here the conventional 'fall-back' value (with drift) has been used, and the example shows that with good process qualities in manufacturing the risk of defective structural properties due to a manufacturing defect can be controlled if combined with fail-safe design and integrity when defects are difficult to detect, and a number of adjacent, load-paths are not affected (in which case an existing, ultimate margin of safety can be used for damage-tolerant critical structure).

5.1.2 *Internal overloads due to manufacturing*

Another set of defects and effects associated with manufacturing involves internal overloads. The results come in three categories:

1. Built-in loads (e.g. poor fastener fits, excessive clamp-up), $\bar{L}_B$; 'Abnormal' load concentrations (e.g. occasional inappropriate fastener, variations in bond-line thicknesses, spot-wise changing part stiffness), $\bar{L}_C$;
2. Spurious geometry (e.g. excessive tolerance, faulty lay-up), $\bar{L}_G$;
3. Internal damage sizes with low detectibility, $\bar{L}_D$.

The contribution to an unsafe state owing to the effects of these types of defect can be written as

$$P(\bar{S}_L \mid \bar{X}_M) = P(\bar{L}_B \mid L_C L_G L_D) + P(\bar{L}_C \mid L_G L_D) + P(\bar{L}_G \mid L_D) + P(\bar{L}_D) \qquad (5.7)$$

An interpretation of orders of magnitude of the defects included in Equation (5.7) is discussed in Example 5.4 below.

Example 5.4 The use of Six-Sigma-type process qualities in producing a starting point for unsafe state probability contributions by manufacturing is assumed. The situations discussed have two difficult features. The basic one is that most of the defects are often very difficult to detect, or are failing without showing any external signs of damage, which makes a fail-safe design a clear option.

The first term on the right-hand side involves spurious loads of an order of magnitude greater than 20% of limit internal load. It represents a type of defect that is hard to detect, except when mechanical damage is present.

A review of Equation (5.7) leads to the conclusion that the structure has to be fail-safe:

$$\begin{aligned}
P(\bar{S}_L \mid \bar{X}_M \bar{Y}_t) &= P(R_t U_T \bar{H}_T U_{T1} H_{T1}) \\
&= P(R_t) \cdot P(U_T \mid R_t) \cdot P(\bar{H}_T \mid U_T) \cdot P(U_{T1} \mid U_T) \cdot P(H_{T1} \mid U_{T1}) \qquad (5.8)
\end{aligned}$$

The special convention for this equation is that a manufacturing defect causes a load-path failure.

An assessment of desirable numerical process values results in

$$\begin{aligned}
P(\bar{S}_L \mid \bar{X}_M \bar{Y}_t) &= 0.99 \cdot 0.90 \cdot 10^{-3} \cdot 0.90 \cdot 0.99 \approx 0.8 \cdot 10^{-3} \\
\Rightarrow P(\bar{S}_L \bar{X}_M \bar{Y}_t) &= 0.8 \cdot 10^{-3} \cdot 0.2 \cdot 10^{-13} \approx 10^{-15} \quad \text{if ten load paths}
\end{aligned}$$

where $t < T$.

Thus the probability of an unsafe contribution can be considered remote for a Seven-Sigma quality.

5.2   DEFECTS IN MAINTENANCE

The maintenance process failures most likely to affect structural performance are connected with:

- scheduled maintenance
- inspections
- repairs.

and the effects are often related to:

- structural property reductions
- internal load increases
- inspection flaws that leave damage undetected.

5.2.1   Defects affecting structural integrity

We return to the definition of the probability of an unsafe state, expressed as:

$$P(\bar{S}_t) = P(\bar{U}_t \bar{H}_t) \tag{5.9}$$

which, extended to elements of safety other than design is based on:

$$P(\bar{U}_t) = P(\bar{B}_t \bar{X}_E T_i V_{ij}) \tag{5.10}$$

Here $\bar{X}_E$ is the defect and is present for the other element

T_i : type of defect
V_{ij} : numerical value of defect type i.

Here the loss of structural integrity, $\bar{U}_t$, can be caused by defects in repair processes:

- damage caused by removal of damaged structure during repair process;
- defects in processing (bonded repairs), e.g. surface preparation;
- defective fastener installations (bolted joints), e.g. reduced strength.

Defects in the scheduled and directed maintenance processes can cause increases in internal loads:

- Inadequate detail upkeep (e.g. jack-screw lubrication) can increase internal loads.
- Missed detection of damage in repair can cause permanent load increases.

Flaws in the inspection can cause reduced structural properties:

- Mistakes in inspection can cause longer flight times with greater damage.
- Defect in inspection processes can result in reduced residual strength and lost fail-safe integrity.

A numerical illustration of effects can be found below in Example 5.5.

Example 5.5   The difference between damage sizes in D_4 and D_5 is evaluated assuming equal length of all the intervals and load and strength both being normally distributed.

Damage in D_5 and B-values results in the following probabilities of survival of n flights, (p_{sn}) and failure in n flights or less p_{fn}:

n	p_{sn}	p_{fn}
1	0.9996	0.0004
100	0.961	0.039
1000	0.67	0.33
3000	0.30	0.70
6000	0.09	0.91

Damage in D_4 and B-values yield the following results:

n	p_{sn}	p_{fn}
1	0.99999941	0.00000059
1000	0.9996	0.0004
3000	0.9988	0.0012
6000	0.9975	0.0025

The difference between limit size damage, D_5, and damage sizes recommended for damage resistance requirements, D_4, is dramatic. This bring us back to the case-by-case considerations for damage requirements, the need for detailed considerations, the safety requirements for damage resistance and the importance of minimizing the probabilities of unsafe states.

The role that effective inspections play in damage-tolerant critical structure and the value of high-quality repair processes cannot be over-emphasized in structural safety. The example above illustrates how damage-tolerance integrity requires B-value residual strength as opposed to mean values.

5.3   OPERATIONAL DEFECTS

The prevalent defects introduced by operations are the result of:

- exceeding limit external loads during and related to flight, causing damage and even loss of load-paths (loss of fail-safe integrity);

- exposure to spurious environmental effects, e.g. turbulence, lightning damage, hail-storms, causing damage or loss of load-path;
- failure to detect and/or report ground damage, often causing loss of damage-tolerance integrity from serious structural damage.

These incidences should be reported immediately to preserve safety and detailed inspections should be conducted to restore integrity. Although the event itself is often short-term, its consequences can be long-term and severe.

Typical threats resulting from these defects are:

- reduced structural properties
- internal overloads
- large mechanical damage sizes.

5.3.1 *Reduced structural properties due to defects in operation*

Defects in operation can result in different kinds of reduction in structural property.

1. They can have long-term effects *when limit loads are temporarily exceeded,* and substantial mechanical damage causes ultimate and/or limit integrity to be lost or fail-safe load-paths to disappear.
2. Exposure to *spurious environments* can cause loss of fail-safe integrity without detection.
3. *Large-scale mechanical damage is not detected or reported* and limit integrity is lost.

The three defects identified above can cause structural property reductions or loss of integrity that can be both long-term and fatal. They can also lead to immediate PSE failure and should be prevented by very high probability avoidance.

5.4 DEFECTS IN REQUIREMENTS FORMULATION

This element of safety involves:

- regulations
- design criteria
- design manuals.

It focuses on requirements affecting structural integrity, the quality of all the elemental processes and the integrity and objectivity of the formulation of regulations.

The basic regulation requirements targets for composite structures are:

- 'Allowables';
- design data;
- mechanical damage sizes;
- demonstration of compliance;
- formulation of required maximum defects sizes;
- demonstration of maximum damage growth rates;
- minimum damage resistance;
- definition of limit loads for critical designs.

Requirements and the state of design practices should be inseparable. They come in many forms in composite structural design, and are mainly non-existent for hybrid structures except for design solutions to the incompatible nature of aluminum/composite contact.

The preliminary philosophy for composites regulations is often quoted as

Regulations for metallic structures often apply directly to composite structures.

However, closer scrutiny shows that commonly there are very distinct differences. The need for separate composites regulations is demonstrated by the following set of examples:

Example 5.6   There are often fundamental differences between metallic and composites allowable values. The metallic design allowable values for transport-category aircraft are, essentially, based on statistical material values corrected to B-value definition but with deterministic environment, deterministic damage for fail-safe and deterministic effects of fastener quality. Residual-strength data are rarely used in the design process, but are a strong factor in determining inspection intervals.

Composite material allowables and design data have strong elements of random descriptions included in their definitions. FAR 25 allows the use of B-values in general when the structure is fail-safe, and the requirements are not material-dependent. Design values for composites are of the type that, if expressed in terms of the probability of 'lost integrity', $\overline{U}$, then

$$P(\overline{U}) = P(\overline{B}_U S_D S_E S_S) = P(\overline{B}_U \mid S_D S_E S_S) \cdot P(S_D) \cdot P(S_E) \cdot P(S_S) \qquad (5.11)$$

where

$\overline{B}_U$, ultimate strength, is less than ULS (ultimate allowable strength), so the first factor is the allowable

S_D : state of damage
S_E : state of the environment (e.g. maximum temperature)
S_S : state of 'local' stress (load) (e.g. open-hole distribution).

A numerical evaluation for a composite's B-value would result in the following probability of losing ultimate integrity:

$$P(\bar{U}) = 0.10 \cdot 10^{-2} \cdot 10^{-3} \cdot 10^{-2} = 10^{-8}$$

while a metallic B-value would result in

$$P(\bar{U}) = P(\bar{B}_U S_D S_E S_S) = 0.10 \cdot 1 \cdot 1 \cdot 1 = 0.10$$

because the states are deterministic in Equation (5.11). Furthermore the state of environment (e.g. maximum temperature) is a temporary state, so in order to violate ultimate integrity it has to coincide with reaching limit load or having an adjacent, failed load path, making this a very rare combination of events. From a book-keeping standpoint a defect of a manufacturing, installation nature, which would create an open-hole situation, would be a very remote possibility.

Residual strength at limit load is often a designing requirement for composite structure, while the 'metal world' uses criteria damage, e.g. failed stringer and a cracked skin-width equal to one stringer spacing in the fail-safe design to assure limit integrity with external limit loads.

Because large-scale damage is a dominant factor in determining the critical design mode for composite structure, a number of damage-tolerance threats must be considered. The following example illustrates a number of threats that it may be necessary to consider.

Example 5.7   We assume that B-values are used for ultimate design and a coefficient of variation, $C_V = 0.10$, for the design data. We also assume a normal distribution. We now ask what the equivalent limit is. The basic definition for ultimate B-values lead to

$$\Phi\left(\frac{B-\mu}{\sigma}\right) = 0.10 \Rightarrow t = -1.3 = \frac{B-\mu}{0.1\mu} \Rightarrow B = 0.87\mu$$

the equivalent value for limit, B_L, is

$$\Phi\left(\frac{\dfrac{0.87\mu}{1.5} - \mu}{0.10\mu}\right) = \Phi(-4.2) = 10^{-5}$$

thus ultimate design with B-values makes 10^{-5} a maximum probability for equivalent limit value.

The resulting requirement for the probability of loss of limit integrity becomes,

$$P(\bar{U}_L \mid \bar{X}) = P(\bar{B}_L D_U \mid \bar{X}) = P(\bar{B}_L \mid \bar{X} D_U) \cdot P(D_U \mid \bar{X}) \qquad (5.12)$$

$$P(\bar{U}_L \mid \bar{X}) \leq 10^{-5} \cdot 10^{-1} = 10^{-6}$$

This requires the probability of loss of limit integrity to be less that 10^{-8}, which aggravates any situation with element defects that have limited detection. A definition interpretation of the probability of an unsafe state, $P(\bar{S})$, can be expressed as considering the detectibility situation:

$$
\begin{aligned}
P(\bar{S}) &= P(\bar{H}_\tau \bar{X}_\tau \bar{B}_{LT} \bar{X}_T D_U \bar{H}_T) \\
&= P(\bar{H}_T \mid D_U \bar{X}_T \bar{B}_{LT}) \cdot P(\bar{B}_{LT} \mid D_U \bar{X}_T) \cdot P(D_U \mid \bar{X}_T) \cdot P(\bar{X}_T) \cdot P(\bar{H}_\tau \mid \bar{X}_\tau) \\
&= 1 \cdot P(\bar{B}_{LT} \mid D_U \bar{X}_T) \cdot 10^{-1} \cdot 10^{-2} \cdot 10^{-2} \cdot 10^{-2} = P(\bar{B}_{LT} \mid D_U \bar{X}_T) \cdot 10^{-7}
\end{aligned}
$$

which for local damage $\leq 10^{-8}$.

Which results in $P(\bar{B}_{LT} \mid D_U \bar{X}_T) \leq 0.10 \Rightarrow$ which requires B-values.

The scenario with impact damage in the interval $\tau \leq t \leq T$ analogously requires $P(\bar{B}_{LT} \mid D_U \bar{X}_T) \leq 0.01$.

The probability of an unsafe state' previously defined leads to a requirement of

$$P(\bar{S}) \leq 10^{-9} \qquad \text{for } \tau \leq t \leq T$$

$P(\bar{S}) = P(\bar{X}_t \bar{U} \bar{H}_T) = P(\bar{H}_T \mid \bar{U}_T \bar{X}_t) \cdot P(\bar{U}_T \mid \bar{X}_t) \cdot P(\bar{X}_t)$ is for limit damage size damage size $D_5 \Rightarrow = 10^{-3} \cdot P(\bar{U}_T \mid \bar{X}_t) \cdot 10^{-2} \leq 10^{-9} \Rightarrow P(\bar{U}_T \mid \bar{X}_t) \leq 10^{-4} \Rightarrow P(\bar{B}_T \mid D_{5T} \bar{X}_t) \leq 10^{-1}$ B-values needed.

Uncertainty in requirements, advice, regulations, practices, criteria and education is a typical situation found in the design of composite structures. Also typical in composite structures are new materials, new processes and new structural concepts and a very slow accumulation of practical service experience and standardization. So the risk, uncertainty and safety of evolving composite structural designs have to be part of a 'cradle-to-grave' activity of monitoring and updating safety levels and reducing uncertainty during the period of the stated service objective. This creates a need for starting positions (the numerical value of the reduction in a structural property due to a process defect), and for design criteria that are case-dependent and that must be initialized and updated in service.

The philosophy adopted in this approach starts with a safety objective that makes the probability of an unsafe state one of the central measures of structural safety, based on safety levels in aluminium structures in transport-category airplanes recognized by the FAA and including the perceived improvements that are needed by the end of the first quarter of this century. Therefore, the case-by-case situation will create a set of numerical requirements based on what will be needed and realistically attainable, and the object of the monitoring.

It will also be a source of objectives for all the elements of safety, and a foundation for process-quality definitions and the links between process defects and the shortfall in structural integrity requirements.

5.4.1   *Regulation targets for composites*

A consistent, safety-based set of design constraints is required to deal with innovation and evolution of composite technology and design.

The present definition of Limit Load (FAR 25)

The largest load expected in service.

must be crisper (i.e. deal with internal loads for design), and provide answers to the following questions:

- Which loads are being dealt with?
- What is the largest possible internal load in flight and on the ground?
- What is the total frequency in a lifetime?
- What is the difference between static and dynamic loads in design (see Figure 5.1)

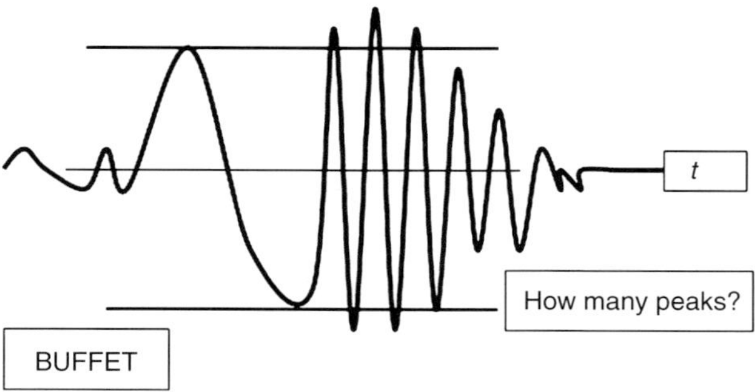

Figure 5.1 Dynamic loads.

(Just to mention a few!)

The present practice for ultimate design values (allowables), is based on an expression such as that the following yielding probability:

$$P(\bar{B}S_D S_E S_S) = P(\bar{B} \mid S_D S_E S_S) \cdot P(S_D) \cdot P(S_E) \cdot P(S_S)$$

where

S_D : state of damage

S_E : state of environment

S_S : state of local load distribution.

- Why is there not a defined state of damage including that for buckling?
- Why is $-65°F$ not required for damaged structure?
- Why is S_S not treated as a defect in manufacturing or repairs with actual probability requirements?

The importance of safe and reliable process quality without drift is a high-level requirement for safe structure in composites. In addition, the demonstration of compliance with an identified target and the demonstration of absence of drift are both necessary regulatory requirements for composites.

5.5 TARGETS FOR MONITORING

An important target for monitoring is to collect damage information in order to update probabilities by use of Bayes' equation (e.g. see Tribus 1969).

$$P(A \mid BC) = P(A \mid C) \left[\frac{P(B \mid AC)}{P(B \mid C)} \right]$$

Here the left-hand side is the probability when we know that both B and C are true.

The first factor on the right-hand side is the probability of A where we only know B.

The second factor shows how the assignment of probabilities changes when we find out about B.

For the more general case when we have many outcomes, the equation is

$$P(A_i \mid BC) = \frac{P(A_i \mid C)P(B \mid A_i C)}{\sum\limits_{j=1}^{n} P(A_j \mid C)P(B \mid A_j C)}$$

This topic will be discussed in detail in Chapter 9.

The focus in monitoring is to review damage size data and analyze the need for updates or modifications to inspection intervals in conformance with risk assessment based on Bayesian updated probability distributions illustrated in Figures 5.2 and 5.3.

Monitoring also includes:

- following up on reported, spurious vibrations to establish the cause;
- conducting walk-around spot-checks to evaluate efficacy;
- collecting data for general damage to test the probability of damage being present, $P(\overline{X})$, test values used in a number of PSEs.

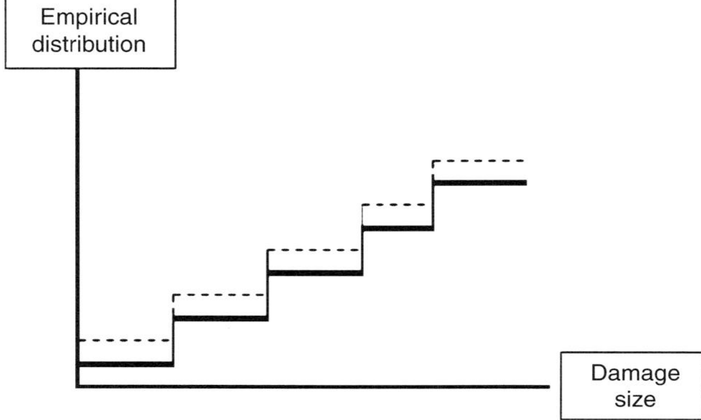

Figure 5.2 Updated distribution.

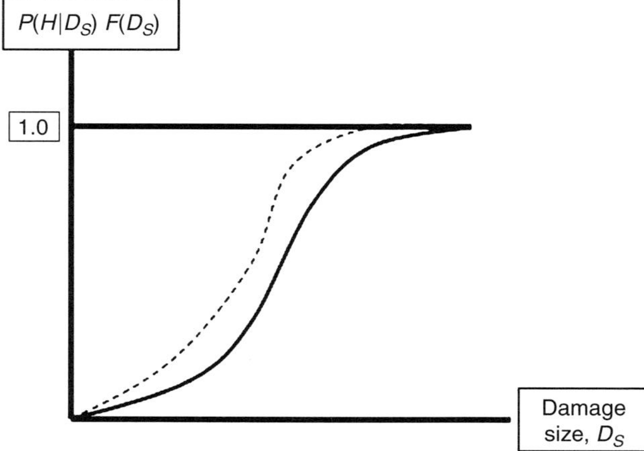

Figure 5.3 Updated distribution.

In Figure. 5.3, the solid line can be used to evaluate the probability of damage distribution given detection (e.g. see Feller 1957).

It is also a very necessary requirement to monitor 'drift in quality of process' in time. The order-of-magnitude requirements demonstrated in this section, and the very difficult task of achieving both these values statistically and yielding reliability, is an issue of the same importance for the structural safety of composites as fatigue is for aluminium structure.

5.6 CONCLUSIONS

Practical safety requirements can result in the need for B-value residual strength.

Initial design values must be based on existing knowledge and supported by a monitoring and updating process based on service observations and inspection results.

Existing regulations and definitions must be updated to meet new composite-specific rules on design data, limit loads and fail-safety.

New rules about the widespread effects of process defects (e.g. for co-bonding) must be formulated/added.

Chapter 6
Design Criteria Development

The purpose of design criteria development is to achieve a balanced set of design rules that identifies the concept of a safe design given, for example, safe manufacturing as a fundamental property. The definition of safe and unsafe manufacturing illustrates the effects resulting from Equation (1.3). Safe manufacturing contains a set of defects that are part of the design decisions and the structural design. The interaction between the elements of safety described in Chapter 4 identified the following defects as common to all elements of safety:

- reduced structural properties;
- excessive internal loads (exceeding Limit);
- spurious mechanical damage sizes.

Chapter 4 also identified defects included in elements of safety and a numerical demonstration that would suggest the following limits for a process with a 'Six-Sigma' quality:

- reduced structural properties $<0.85 \cdot$ allowable values for structural properties;
- excessive internal limit loads $<1.15 \cdot$ defined values of limit internal loads;
- excessive damage sizes larger than sizes in region D_5.

Section 6.1 describes the design decisions that are affected by the moderate defects which are part of safe manufacturing. The other effects discussed in Section 6.3, which deals with unsafe manufacturing and the action necessary to keep probabilities and process-control deviations within bounds, are represented by the term 'probability of unsafe manufacturing'.

This chapter reviews all of the elements of safety in terms of defects.

6.1 FOUNDATION OF SAFETY-BASED CRITERIA

The quality of processes, as earlier described, will yield tolerable defects for elements. Example 6.1 illustrates the effects of both defects of the processes and the variation of Structural integrity (residual strength), with or without detectable damage sizes, all of which provide important inputs to design criteria.

The most troublesome scenario involves loss of structural integrity with undetectable damage. Example 6.1 illustrates the situation when ultimate design is performed with allowable B-values but manufacturing defects cause property reductions.

Figure 6.1 shows variation for $0.9\,\mu$ and $0.8\,\mu$. As labeled, the figure represents a range of strength reductions. The converse can be deduced for internal loads by following a similar argument.

Example 6.1 Normal distribution is assumed, the equation for B-values yielding

$$\Phi\left(\frac{B - \mu}{C_V\mu}\right) = 0.10 \Rightarrow B = 0.87\,\mu$$

and k is strength reduction

$$\Phi\left(\frac{B - k\mu}{k\mu C_V}\right) = 0.10 \Rightarrow B = 0.87\,k\mu$$

$$10\%\ \text{reduction} \Rightarrow k = 0.9 \Rightarrow 0.78\,\mu$$
$$20\%\ \text{reduction} \Rightarrow k = 0.8 \Rightarrow 0.70\,\mu$$

C_V is assumed to be 0.1. Figure 6.1 illustrates the resulting B-values for 10% and 20% strength reductions, and, as this case focuses on the situation when the probability of detection is zero, the fail-safe design approach must be used but can only be realized when the defective process only affects one load-path; otherwise, fail-safe integrity does not exist.

The total effects of all of the elements of safety have, from the first analysis given in Chapter 2, shown that a very common result of, for example, manufacturing defects, is

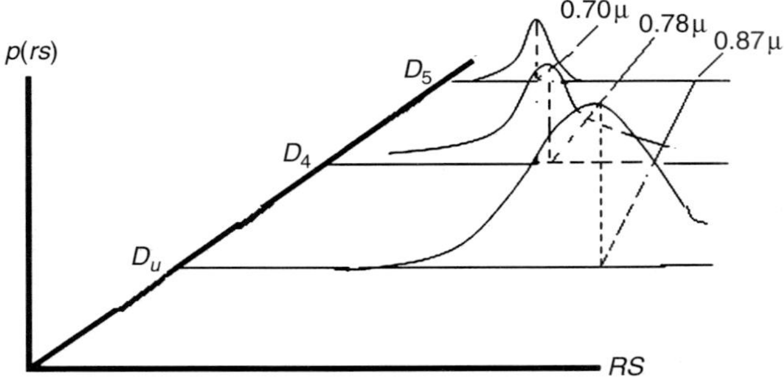

Figure 6.1 Residual strength distributions for different damage sizes.

the occurrence of property reduction without detectable damage. The interaction between all of the elements also produces scenarios for which the requirements, even though ultimately relatable to PSEs, have critical value, for example for process failures, dependent on the requirements for one or some of the sub-regions of a PSE as opposed to design requirements that can depend on all of the DDPs.

The consequence of these case-by-case design algorithms requires a detailed knowledge of the causes of the different defects and how they interact through the design models, and how they also have a framework for interaction of different defects and elements of safety that makes it possible to produce unified interaction criteria.

Example 6.2 will be used to demonstrate how the reduction of strength affects the probability of survival.

Example 6.2 We assume that load and strength are normally distributed and that the mean shows a 20% reduction of strength and is limited to a local area:

$$\mu_X \ : \ 0.7 \, LLS \text{ and } \mu_Y = 1.15 \, LLS \cdot 0.8$$
$$\mu_G \ : \ 0.92 \, LLS - 0.7 \, LLS = 0.22 \, LLS$$
$$\sigma_G \ : \ C_V \, LLS \, (0.85 + 0.49)^{.5} = 0.16 \, LLS$$
$$t \ \ \ : \ -1.38 \Rightarrow \Phi(-1.38) = 0.084$$

yielding a probability of surviving any flight of 0.916.

A reduction of 20% in strength results in the following probabilities of survival:

n	p_S	Probability of failure after n flights or fewer
1	0.916	0.084
10	0.42	0.58
100	0.0002	0.9998
200	~0	~1.0000

The result again reinforces the need to design fail-safe structure for these types of PSE.

A reduction of 10% in strength yields

$$\mu_G \ : \ 1.035 \, LLS - 0.7 \, LLS = 0.335 \, LLS$$
$$\sigma_G \ : \ C_V \, LLS \, (1.071 + 0.49)^{0.5} = 0.125 \, LLS$$
$$t \ \ \ : \ -2.68 \Rightarrow \Phi(-2.68) = 0.0044 \Rightarrow p_s = 0.9956$$

n	p_S	p_{Fn}
1	0.9956	0.0044
100	0.64	0.36
500	0.11	0.89
1000	0.01	0.99

These results again support fail-safe design. It is interesting to evaluate the probability of being below limit strength when ultimate design is based on ultimate B-values.

The definition of B-values yields

$$\Phi\left(\frac{x-\mu}{C_V\mu}\right) = 0.10 \Rightarrow \frac{x}{\mu} = 1 - 1.3 \cdot C_V$$

which yields the probability of strength being less than limit internal load:

$$\Phi\left(\frac{\dfrac{1}{1.5}(1-1.3\,C_V)-1}{C_V}\right) = \Phi(t)$$

C_V	$\dfrac{x}{\mu}$	T	$\Phi(t)$
0.10	0.87	-4.2	$1.3 \cdot 10^{-5}$
0.09	0.883	-4.6	$0.2 \cdot 10^{-5}$
0.08	0.896	-5.03	$0.3 \cdot 10^{-6}$

The C_V-range shown is often a 'practical' range for structural design values and yields an average upper bound of $\sim 10^{-5}$ to 10^{-6}. Thus, when there is no detectable damage present, this range of values seems to make fail-safe design necessary.

The safety-based, structural design consequence in the presence of unreliable detection is fail-safety. A comparison with the results in Chapter 2 shows the advantage of fail-safety, especially when detection opportunities are limited.

6.2 FAIL-SAFETY, STRUCTURAL DESIGN AND SAFE MANUFACTURING

Fail-safe designs in composite and hybrid structures are much more complicated to produce and validate than in metallic structures, in particular because of the need to demonstrate a capability for load redistribution after load-path failures, especially in compression. A typical fail-safe scenario for a PSE can be expressed as

$$P(\bar{U}_t \bar{Y}_t R_t S_{UT} \bar{H}_T) = P(\bar{S}_T) \tag{6.1}$$

Here:

> a critical load path has lost limit integrity, $\bar{U}_t$ at or before t
> the focus load path fails, $\bar{Y}_t$, at t the internal loads are redistributed R_t
> internal fail-safe integrity of the remaining structure, exists, S_{UT}
> the load path failure is not detected at major inspection at T

$$P(\bar{U}_t) = P(\bar{X}_M T_1 V_{12})$$

Here the integrity is lost due to manufacturing defects; strength is lost due to processing error.

It is clear that the results of processing errors can afflict all the Design Detail States (DDSs), causing a stochastic dependence. The probability of a safe state for a PSE, PSE_1, can be expressed as

$$\begin{aligned} P(PSE_1) &= P(DDS_{11} \cdot DDS_{12} \cdots DDS_{1n}) \\ &= P(DDS_{11} \mid DDS_{12} \cdots DDS_{1n}) \cdots P(DDS_n) \\ &= 1 \cdots 1 \cdot P(DDS_n) = P(DDS_n) \end{aligned}$$

Independence between the resulting damage sites yields:

$$\begin{aligned} P(PSE_1) &= P(DDS_{11}) \cdots P(DDS_{1n}) \quad \text{if equal} \\ &= [P(DDS_1)]^n \Rightarrow P(\overline{PSE}_1) \approx n \cdot P(\overline{DDS}_1) \end{aligned}$$

A thorough understanding of the nature of the processing is required to determine safety targets, which can differ by a factor of n.

In continuation, the investigation of an unsafe state can be achieved with the following expansion of Equation (6.1):

$$\begin{aligned} P(\bar{S}_T) = {}& P(\bar{H}_T \mid \bar{U}_t \bar{Y}_t R_t S_{UT}) \cdot P(S_{UT} \mid \bar{U}_t \bar{Y}_t R_t) \cdot P(R_t \mid \bar{Y}_t \bar{U}_t) \cdot P(\bar{Y}_t \mid \bar{U}_t) \\ & \cdot P(V_{12} \mid T_1 \bar{X}_M) \cdot P(T_1 \mid \bar{X}_M) \cdot P(\bar{X}_M) \end{aligned} \tag{6.2}$$

Here:

> the first factor is an undetected 'failed load path' at a major inspection at T
> the second factor represents the probability of a 'safe state' for the remaining DDPs
> the third factor is the probability of a successful redistribution of internal loads after a load path failure
> the fourth factor is the probability of a specified value of a manufacturing defect, (e.g. 20% strength reduction)
> the fifth factor is the probability of a specified manufacturing flaw
> the sixth factor is the probability of a manufacturing defect.

This scenario is very important for structural safety in a situation when detection is unlikely. The next example provides a numerical interpretation of the probability of an unsafe state.

Example 6.3 This example is based on the presumed data required to produce a safe structure, the purpose being an order-of-magnitude view of safety requirements. The tool for this evaluation is Equation (6.2).

$$P(\overline{S}_T) = 10^{-3} \cdot 0.90 \cdot 0.90 \cdot 10^{-1} \cdot 10^{-1} \cdot 10^{-1} \cdot 0.33 \cdot 3.4 \cdot 10^{-6} = 0.9 \cdot 10^{-12}$$

It should be noted that this value is based on a 'drift' of 1.5σ, while the true Six-Sigma probability value is considerably lower, yielding an adequate 'improvement cushion'.

There are many reasons for making fail-safe designs, but elimination of processing defects of poor detectability is the most persuasive, requiring close attention to the detailed processing requirements and the structural consequences due to defects in manufacturing and other elements of safety.

It has been noted repeatedly that, to succeed in load redistribution in composite structure, preserved ultimate integrity its also involved which is most effectively solved by requiring a 20% margin of safety for ultimately critical structure, making limit damage sizes safer. Many of these concerns are case-to-case dependent and the approach formulates principles and philosophies to be applied to individual examples of composite and hybrid structures.

6.3 REDUCED STRUCTURAL PROPERTIES AND UNSAFE MANUFACTURING

Potential process defects in manufacturing processes must be controlled to 'designed-in' levels. The quality of process characteristics is a case-by-case requirement and the baseline is the probability of an unsafe state which is shared by the elements of safety. A natural unit is PSE when formulating requirements, but units as sub-assemblies are just as realistic, provided the number of units of the vehicle can be kept within a practical range.

We return to a property reduction of 30% which, under the assumption of normal distribution, yields a probability for surviving a random flight of 0.933. Surviving n consecutive flights is associated with the following probabilities:

n flights	Probability of surviving n consecutive flights	Probability of failing in n flights or fewer
1	0.933	0.067
100	10^{-3}	0.999
1000	~ 0	~ 1

The 30% reduction in strength implies that the probability of being below failure strength is 0.99 and fail-safe integrity is lost. This result is in an intuitively unsafe state. We return to the definition of an undetected loss of limit integrity,

$$P(\bar{S}_T) = P(\bar{U}_T\bar{H}_T) = 1 \cdot P(\bar{U}_T)$$
$$= P(\bar{B}_T \mid \bar{X}_M T_1 V_{1US}) \cdot P(V_{1US} \mid T_1\bar{X}_M) \cdot P(T_1 \mid \bar{X}_M) \cdot P(\bar{X}_M) \qquad (6.3)$$

V_{1US} is an unsafe defect value in processing.

In the next example is illustrated the numerical consequences of large, unsafe defect values. Equation (6.3) is evaluated as implicit, assuming undetectable defects.

Example 6.4 We assume that Six-Sigma is a natural starting point. The strength is assumed to be reduced by 30%. It is normally distributed with $C_V = 0.10$, leading to the probability:

$$\Phi\left(\frac{LLS - 0.7 \cdot 1.15 \cdot LLS}{C_V \cdot 0.7 \cdot 1.15 \cdot LLS}\right) = \Phi\left(\frac{0.24}{0.10}\right) = 0.9918$$

Equation (6.3) then produces the following probability for an unsafe state:

$$P(\bar{S}_T) = 0.99 \cdot 10^{-3} \cdot 0.5 \cdot P(\bar{X}_M) = 0.5 \cdot 10^{-3} \cdot P(\bar{X}_M)$$

If we generalize Six-Sigma to N-Sigma we get the following results:

N	$\Phi(-N)$	Reg. for an unsafe state	Prob. of unsafe defect
4.5	$3.4 \cdot 10^{-6}$	10^{-13}	$2 \cdot 10^{-10}$
6	10^{-9}	10^{-13}	$2 \cdot 10^{-10}$
~ 7			
8	$6 \cdot 10^{-16}$	10^{-13}	$2 \cdot 10^{-10}$

This table shows a very difficult process quality requirement of ~ 7, with the exception of a long-term goal with a gradual transition in defects between 6σ and 7σ, not to speak of data requirements for monitoring for updating.

The only 'reasonable' choice seems to be to assure a margin of safety in design that preserves fail-safe integrity. The following chapter will review the limits of what defects are acceptable, what must be prevented, process control and quality control. Six-Sigma quality must cover up to $\sim 15\%$, dependent on the relations between processing defects and structural integrity levels.

6.4 CONCLUSIONS

A very close review of process quality requirements shows that the principle of Six-Sigma quality is useful in ensuring safety in innovation.

The specifics must be developed for an *N*-Sigma requirement and used on a PSE basis.

The resulting criteria can be based on the effect of the different defects and apply to the characteristics of specific PSE's.

Chapter 7
Scenarios and Structural Safety

Structural safety is improved by damage detection. Large-scale damage very often is easily detectable. The threat to safe states can often be mitigated by detection. Damage locations that are accessible to 'walk-around' inspections have extra safety protection. There are, however, a series of situations where neither situation is at hand. A safe state can turn to unsafe just after a major inspection, and practical inspection periods are not survivable. Walk-around inspections may not be possible because of poor access. Defects in, for example, manufacturing, can cause loss of integrity without any sign of damage.

It has already been shown in Backman (2005) that an unsafe state leads to failure in a practically short time, if the damage tolerance requirements are based on large-scale damage, and the structural solution becomes fail-safety.

The probability of an unsafe state between inspections can be expressed as

$$P(\bar{S}_t) = P(\bar{S}_T \bar{A}_t) + P(\bar{S}_T A_t) \rightarrow P(\bar{S}_T A_t)$$

when the number of flights increases and $P(\bar{S}_T A_t) = P(A_t \mid \bar{S}_T)P(\bar{S}_T)$

Here:

A is survival in n flights and S is a safe state

An analogous argument yields

$$P(\bar{A}_n) = P(\bar{A}_n \mid \bar{S}_T)P(\bar{S}_T) = 1 - P(A_n)$$

which for a scenario of 'no growth' or for the region where residual strength is insensitive to damage size

$$P(A_n) = p_s^n \cdot P(S_T)$$

where p_s is probability of survival.

The scenario described here deals with mechanical damage and the damage tolerance situation in general. The probability of surviving n flights while in an unsafe state is

n	$P(A_n)$	Probability of failing after at most n flights
1	0.998	0.002
100	0.82	0.18
1000	0.14	0.86
3000	0.003	0.997

and deals with a numerical situation that could represent damage in region D_5. The probability of failure is not encouraging without walk-around inspections.

7.1 CRITICAL SCENARIOS

The structural design scenarios which often include mechanical damage result in the choices of structural properties, damage size requirements, and inspection approaches and periods. Both design decisions and risk management are focused on producing, maintaining and restoring structural safety. In the preservation of damage tolerance, structural integrity is the centerpiece of structural safety. However, fail-safe design and walk-around inspections have, wherever possible, developed into the saving grace for maintaining safety when flaws are hidden or non-detectable.

The cornerstones of structural safety have become structural integrity, stringent, scheduled inspections, fail-safe design and walk-around inspections.

An overall analysis of the effects of other elements of safety has reinforced the importance of these cornerstones. The types of defects that are involved are:

- Reduction in structural properties due to defects in processing (that do not cause mechanical damage), hidden defects in repairs or defects in regulations, etc.;
- Local increases in internal loads caused by geometric deviations, clam-up and incorrect fastener choices, increased activation loads due to flaws in scheduled maintenance.

Both these effects can, in time, lead to the development of mechanical damage and local load path failures.

These effects and their threats to safety require fail-safe design and walk-around inspections wherever possible.

The probability of defects for different processes present with different elements of safety. It is a very important consideration and can be expressed as

$$P(\overline{X}_S) = P(\overline{X}_M T_I V_{IJ}) = P(V_{IJ} \mid T_I \overline{X}_M) \cdot P(T_I \mid \overline{X}_M) \cdot P(\overline{X}_M) \tag{7.1}$$

S : the subscript indicating process, $S = M$; M is manufacturing
$(\overline{X}_M)$: a defect in manufacturing
T_I : the type of defect
V_{IJ} : the numerical value; e.g. if equal to 0.8 it means that only 80% of a structural property remains.

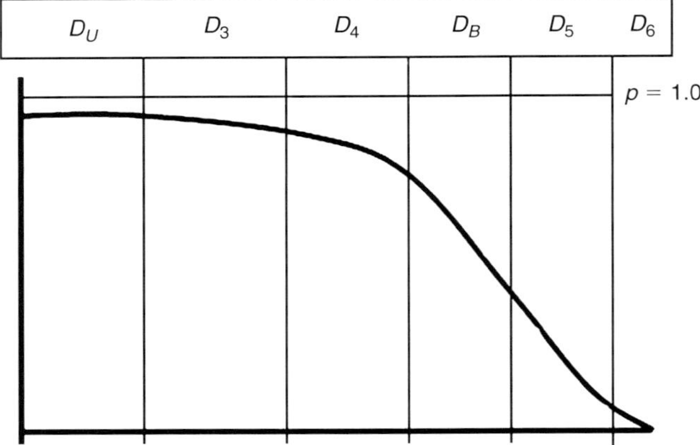

Figure 7.1 Damage sizes versus survival.

The first factor on the right-hand side of Equation (7.1) must be limited to a worst case scenario, belonging in this case under 'safe manufacturing'.

For Six-Sigma processes it is often found that reasonable values to suit the requirements cannot be worse than a 20% reduction of structural properties or a 20% increase in maximum internal load levels.

Integral to safety scenarios are probabilities of survival of n flights with large damage, e.g. an inspection period. Figure 7.1 shows a desirable arrangement of damage size regions. Here D_U is the ultimate design region; D_3 is a transition region; D_4 is the damage resistance region; D_B is a growth zone; and D_5 is the limit region. D_4 is a region where good probability of survival can be expected. Survival in region D_5 determines inspection period lengths.

A numerical of example of probability of survival in region D_4 is shown below.

Example 7.1   We assume equal length regions, normal distributions, and B-values in D_5. The limit load stress, *LLS*, is $0.87\,\mu$ for D_5 ($C_V = 0.10$). The mean for damage in $D_4 = 1.48\,LLS$ and the mean for internal loads is $0.7\,LLS$. The normal distribution function yields the probability of failure

$$\Phi\left(\frac{-0.78LLS}{0.16LLS}\right) = \Phi(-4.87) = 0.59 \cdot 10^{-6}$$

The probability of surviving 3000 flights with a damage size in D_4 is 0.998.

The first two scenarios are now shown for reference, describing damage sizes during inspection intervals. The first one deals with growth (see Figure 7.2).

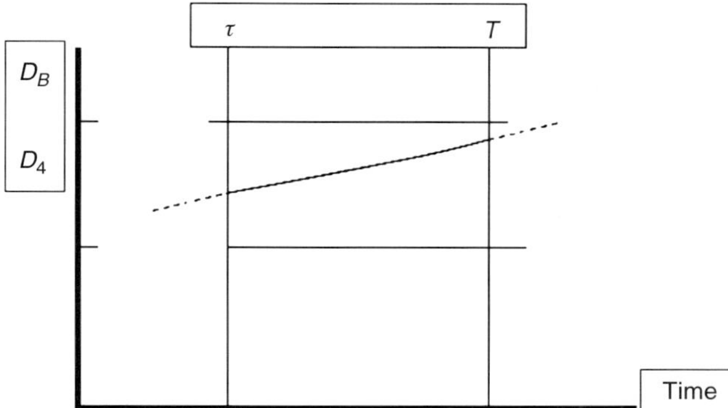

Figure 7.2 Scenario 1: damage size in inspection interval.

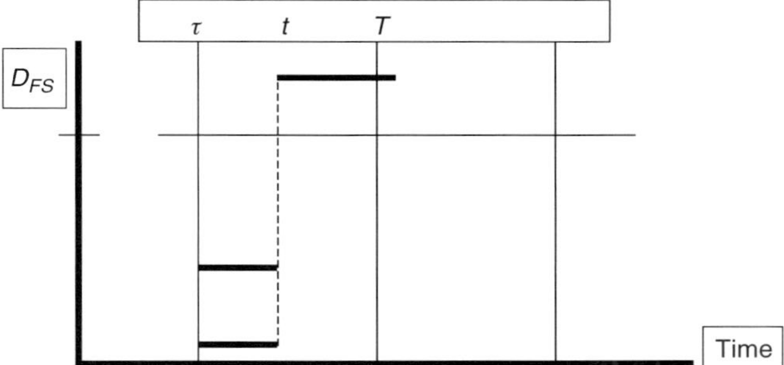

Figure 7.3 Scenario 2: fail-safe for several initial damage sizes.

The second scenario represents a fail-safe case with a failed load-path, where region D_{FS} represents the size for a load-path failure when there are no defects in the other adjacent load-paths and successful load redistribution occurs (see Figure 7.3).

The third scenario deals with widespread manufacturing and repair defects that lead to reduced structural properties that are often hard to detect. These result in loss of structural properties that are not limited to one load-path, e.g. in curing and co-bonding of assemblies and sub-assemblies (e.g. wing and stabilizer surfaces), where fail-safe integrity is violated.

The remedy for these situations could be the required innate, ultimate margin of safety that results from damage tolerant criticality. That would allow us the following approach:

Example 7.2 We assume damage-tolerant critical structure, which lead to the following equivalent thickness based on internal design load, N_U:

$$t = \frac{N_U}{1.5F_L} \text{ leading to an ultimate margin of safety}$$

$$MS_U = \frac{F_U t}{N_U} - 1, \text{ now we assume an ultimate reduction, } `k' \Rightarrow$$

$$1 + MS_U = \frac{F_U}{1.5 F_L} \text{ if we now reduce by } `k' \text{ so that } MS_U = 0 \Rightarrow$$

$$k = 0.8 \Rightarrow MS_U = +25\%$$

So the limit maximum damages D_5 could be increased $[1.25/1.20]^2 = 8.5\%$ with a square-root assumption for residual strength.

If, however, the residual strength properties are also reduced by the manufacturing defects, the conclusion becomes very different, e.g. reduced damage size. So again the knowledge of how the specific process and monitoring behave is a very important part of the design criteria and data and therefore need to be corrected when detected.

So the quality is defined as

$$\Phi\left(\frac{\mu - 6\sigma - \mu}{\sigma}\right) = \Phi(-6)$$

which is interpreted as 1 defect in 10^9 realizations and with a drift of 1.5σ 3.4 defects in 10^6 realizations. This short-term quality, as shown in the summary and even though often quoted in process development, will be rejected in this context.

The question is now: 'How does this protect the allowable values that we write under the normal assumption?'

7.2   CONCLUSIONS

A practical set of safety rules in design must be derived from a realistic damage scenario and applied to PSEs for specifics.

Walk-around inspections, applied whenever possible, are very effective supports to structural safety.

Adaptation of Six-Sigma to different scenarios, locations, and situations very often deals highly effectively with safety.

Chapter 8
Safety and Structural Integrity

The review of structural safety in previous chapters has identified the nature of safety threats emanating from different elements of safety. The role of structural integrity in the quest for safety has been highlighted. We now describe the interaction involved.

8.1   ULTIMATE INTEGRITY, THE BASIC CORNERSTONE

The basics involve the interaction of ultimate integrity with other types of integrities such as fail-safe integrity of the design, which is a justification for B-values. The cornerstones of ultimate integrity therefore are a safety factor of 1.5 and B-values for 'allowables' and design data, which also protect fail-safe integrity and the basic limit integrity. The definition of B-values leads us to the following equation for ultimate B-value, F_U, which with practical, reasonable scatter in the ultimate design data leads to the following minimum limit integrity requirement:

Example 8.1   We assume a normal distribution, and $C_V = 0.1$.

$$\Phi\left(\frac{F_U - \mu}{\sigma}\right) = \Phi\left(\frac{\dfrac{F_U}{\mu} - 1}{C_V}\right) = 0.10 \Rightarrow t = -1.3 \Rightarrow \frac{F_U}{\mu} = 0.87 \Rightarrow \text{implies the equivalent}$$

$$\text{probability of limit strength } \Phi\left(\frac{\dfrac{0.87\mu}{1.5} - \mu}{C_V \mu}\right) = P\left(\frac{\dfrac{0.87}{1.5} - 1}{C_V}\right) = P(-4.2) = 10^{-5} \text{ so for}$$

this case ultimate B-value strength protects limit strength at a 10^{-5} probability, the significance of which will be explained later, but certainly raises concern about limit integrity.

8.2   PROCESS QUALITY AND MONITORING

Within the numerical illustrations in previous chapters the baseline objective of Six-Sigma quality and the associated 'drift' and how this protects structural integrity have come up often; these considerations are very important in terms of maintaining structural safety. It is common to define process quality in terms of probability of defects with a 'drift' of 1.5. Feller (1957) shows that the sum of defects in time, $X(t)$, is a Poisson process defined by the Poisson distribution identified in Equation (8.1):

$$p_x(x) = \frac{(\lambda t)^x e^{-\lambda t}}{x!} \quad x = 1,\ 2,\ 3..., \tag{8.1}$$

The stochastic process $X(t)$ is the number of defects since time $t = 0$. Observations of the random process $X(t)$, $x_1(t)$, $x_2(t)$, ... count the total number of defects in time. It also shows that the exponential distribution

$$f_T(t) = \lambda e^{-\lambda t} \tag{8.2}$$

describes the times between consecutive defects; and λ is the average time between events (defects). Establishing and maintaining process quality can therefore be influenced by both cost and safety.

Before proceeding with an analysis of specific defects, we will investigate the relation between defects and repair or replacement. Slooman (1968) makes the connection and shows that the density function of renewal times for n renewals is

$$f(t) = \frac{\lambda (\lambda t)^{n-1} e^{-\lambda t}}{(n-1)!}$$

which is a Poisson process and the time to first renewal is $t = 1/\lambda$. The setting of safety requirements can therefore be combined with safety targets.

The probability that the time to $(n + 1)$, i.e. the arrival of a defect greater than t is: $e^{-\lambda t}$ with a mean $\mu = 1/\lambda$ and a standard deviation $\sigma = 1/\lambda$; the Six-Sigma interval upper limit is

$$t \leq \frac{1}{\lambda} + 6\frac{1}{\lambda} = \frac{7}{\lambda} \Rightarrow e^{-\lambda t} = e^{-7} = 0.9 \cdot 10^{-4}$$

The probability that n defects will be contained in the six-sigma interval at time t

$$P(NT) = P(N \mid T \mid \cdot P(T) = P\{N = n\ given\ T\} \cdot P\{T = t\}) = \frac{(\lambda t)^n e^{-\lambda t}}{n!} \cdot e^{-7}$$

which is only dependent on one parameter, calculated from the certification data or from the quality and cost requirements of $P(NT)$ and the renewal requirements defined above.

8.3 DEFECT TARGETS AND RELATIONS

To evaluate defects and set safety targets for each element of safety and their interaction it is necessary to produce safety-based design constraints and process quality necessary to satisfy safety objectives.

8.3.1 *Defects in the design process*

Defects introduced by errors in design causing mechanical damage in time, and the defects produced by the other elements of safety are included here. These potential defects must be based on the definitions of the contributions to the probability of an unsafe state:

$$\Delta P(\bar{S}_T) = P(\bar{S}_D \mid S_R) \tag{8.3}$$

There has been much attention paid to safety levels in establishing and preserving structural integrity in the presence of flaws and damage from external (as opposed to design) sources. The process quality, which is a large part of safety has also been mentioned here. The quality of the 'non-design' elements of safety has been discussed in previous sections.

The quality of the design process and the defects involved are now introduced. The defects in the design process that can be of great harm are those that are not caught in the testing phase and during the 'demonstration of compliance' in the certification process, but which evolve in service. Some examples are:

- weaknesses that develop into damage, because loads resulting from scheduled maintenance actions, inspections and repairs have not been included in the design;
- incorrect internal design loads due to misinterpretation of FE results;
- inadequate-quality design allowable data due to mistakes in establishing critical load cases, or critical combinations of environments and load cases;
- neglecting critical requirements not spelt out in the regulations.
- failure to recognize that the content of this list is dependent on choice of composite material and process and selection of structural concepts, and that its overall thrust will change with any variation in the individual items therein.

The quality of the design process is examined in the following example.

Example 8.2 We start with an illustration of Six-Sigma applied to ultimate integrity, and consequently to loss of fail-safe integrity, which, in itself, is an unsafe state.

$$P(\bar{S}_\tau) = P(\bar{X}_D T_1 V_1 \bar{X}_\tau \bar{H}_\tau D_{3\tau} \bar{B}_{U\tau}) = P(V_1 \mid T_1 \bar{X}_D) \cdot P(T_1 \mid \bar{X}_D) \cdot P(\bar{X}_D) \cdot P(\bar{H}_\tau \mid D_{3\tau})$$
$$\cdot P(\bar{B}_{U\tau} \mid D_{3\tau}) \cdot P(D_{3\tau} \mid V_1)$$

which yields

$$P(\overline{S}_\tau) = 0.5 \cdot 0.10 \cdot 3.4 \cdot 10^{-6} \cdot 0.5 \cdot 1 \cdot 1 = 0.85 \cdot 10^{-7}$$

which makes either the quality of the design process inadequate or the 'drift' too large.
  Here:

V_i :   the measure of the defects, and V_3 means will be found
T_i :   type of defect
$\overline{X}_D$:   design process flaw present
$\overline{X}_t$:   mechanical damage present at t
$\overline{H}_t$:   damage not detected at t
D_{3t} :   damage size at t
$\overline{B}_{Ut}$:   strength is less than design ultimate allowable.

Example 8.2, if nothing else, shows that the quality of the process is very important for safety, but it also supports the following expression:

$$P(\overline{S}_D \,|\, S_R) = P[(\overline{S}_D S_{DP} \,|\, S_R) \cup (\overline{S}_D \overline{S}_{DP} \,|\, S_R)]$$

$$= P(\overline{S}_D \,|\, S_{DP} S_R) \cdot P(S_{DP} \,|\, S_R) + P(\overline{S}_D \,|\, \overline{S}_{DP} S_R) \cdot P(\overline{S}_{DP} \,|\, S_R) \quad (8.4)$$

Where the following events participate:

S_D :   safe design
S_{DP} :   safe design process
S_R :   safe requirements.

Thus, the probability of an unsafe state consists of two parts. One deals with design process unrelated defects and the other, design process failures.
  The probability of an unsafe state caused by 'an undetected loss of damage tolerance integrity' is the subject of Backman (2005). This applies to damage tolerance and often turns out to be critical for composite designs and should be well managed. Its order of magnitude depends on risk management, uncertainty mitigation and monitoring to be included in the safety management. A specific number of PSE's and DDP's, considered typical, are involved

$$\Delta P(\overline{S}) = P(\overline{H}_\tau \overline{U}_T \overline{H}_T) = 10^{-2} \cdot 10^{-6} \cdot 10^{-3} = 10^{-11}$$

The next section deals with manufacturing defects.

8.3.1.1 Defects caused by manufacturing

Effects caused by manufacturing defects are grouped in the following three categories:

1. Reduced structural properties
2. Added internal loads
3. Mechanical damage produced by accident.

The quality of this process can be evaluated from the standpoint of cost. The next example describes a starting-point for establishing process quality.

Example 8.3 Equation 8.1 is used to illustrate the probability of the number of defects in time. Assuming that the target for acceptable number of defects in 10 hours is ten, then the probability of that event is

$$\Pr\{N = 10\,/\,10 \text{ hours}\} \text{ is } \frac{10^{10} \cdot e^{-10}}{10!} = 0.13$$

The understanding of repair cost and the cost of airplanes being taken out of service would then allow us to improve the quality from the standpoint of both cost and safety.

Example 8.4 We will now investigate what we should include of the possible nature of defects in the safety targets, beginning with reduced structural properties.

$$\Delta P(\bar{S}_M) = \Delta P(\bar{S}_{M1} + \cdots + \Delta P(S_{Mn}))$$

We will now address the *n* properties. These are:

- reduced strength and reduced stiffness (causing reduced buckling 'strength');
- reduced damage resistance;
- increased maximum damage growth rate.

because they are the main contributors to the unsafe state. Example 8.2 illustrates the design process, and the effect on ultimate integrity by the three compromised properties above will now be considered.

We consider the expected value of the detectable measure of the defects.

The effect of the defect that produces reduced strength is

$$\Delta P(\bar{S}_T) = P(\bar{X}_M T_1 V_{13} \bar{H}_0 \bar{B}_U) = P(\bar{B}_U \mid \bar{X}_M T_1 V_{13} \bar{H}_0) \cdot P(\bar{H}_0 \mid \bar{X}_M T_1 V_{13})$$
$$\cdot P(V_{13} \mid T_1 \bar{X}_M) \cdot P(T_1 \mid \bar{X}_M) \cdot P(\bar{X}_M) \tag{8.5}$$

The first evaluation involves a B-value reduced to 70% of its intended ultimate value, *ULS*:

Assuming a normal distribution with a C_V of 0.10

$$\Phi\left(\frac{ULS - 0.70\ ULS}{0.10 \cdot ULS \cdot 0.70}\right) = 1 - \Phi(-3.0) \approx 1.0$$

Equation (8.5) yields

$$\Delta P(\overline{S}_T) = 1 \cdot 1 \cdot 0.10 \cdot 0.10 \cdot 10^{-6} = 10^{-8}$$

The properties using mean values, e.g. E, have a contribution to the total that is approximately the same order of magnitude; i.e. the effect of the first two properties above is

$$2 \cdot 10^{-8}$$

Example 8.5 The result of the increase in the maximum damage growth rates relating to property, especially when it is caused by process failure of composite components such as sub-assemblies causing a widespread increase of growth rate, is presented in this example.

The contribution to an unsafe state can be described as

$$\Delta P(\overline{S}_T) = P_3(\overline{X}_{MG} D_{Ut} D_{3T} \overline{B}_{UT} \overline{H}_{3T}) = P(\overline{H}_T \mid D_{3T}) \cdot P(\overline{B}_{UT} \mid D_{3T})$$
$$\cdot P(D_{3T} \mid D_{Ut} \overline{X}_{MG}) \cdot P(D_{Ut} \mid \overline{X}_{MG}) \cdot P(\overline{X}_{MG}) \cdot \binom{10}{3} \tag{8.6}$$

Here:

3 refers to three loads

$\overline{X}_{MG}$: manufacturing defect that causes increased damage growth
D_{Ut} : Damage at time t in size interval D_U (ultimate damage size).

We now assume that damage of size D_3 violates ultimate integrity and the order of magnitude of the resulting probability is

$$\Delta P(\overline{S}_T) = 0.5 \cdot 10^{-2} \cdot 0.5 \cdot 0.5 \cdot P(V_3 \mid T_3 \overline{X}_M) \cdot P(T_3 \mid \overline{X}_M) \cdot P(\overline{X}_M) \cdot 120$$
$$= 15 \cdot 10^{-2} \cdot 0.1 \cdot 0.33 \cdot 10^{-6} = 0.5 \cdot 10^{-8}$$

The contribution to an unsafe state due to ultimate integrity violation is of the order of magnitude of

$$\sim 10^{-8}$$

if we assume that the marginal probability for each property 'reduction' is 0.33.

8.3.1.2 Defects caused by maintenance

Effects of defects caused by maintenance are:

1. reduced structural properties due to repair;
2. added internal, spurious load due to repair;
3. large damage not detected (by oversight in inspection);
4. 'missed' scheduled maintenance (e.g. lubrication), causing local damage;
5. mechanical damage inflicted in scheduled maintenance (e.g. stepping damage) and repair (e.g. damage at clean-up or removal).

The reduced structural properties due to defects in maintenance produce the same contribution to the probability of an unsafe state as analogous defects produced in manufacturing, ergo 10^{-8}.

Added internal, spurious loads can be the results of defects in the installation of fasteners or bonding of the repair. The probability of loss of ultimate integrity is:

$$\Delta P(\overline{S}_T) = P(\overline{X}_\tau H_\tau R_\tau \overline{X}_{It} T_2 V_{23} \overline{B}_{Ut} \overline{H}_\tau) = P(H_\tau \mid \overline{X}_\tau) \cdot P(\overline{X}_\tau) \cdot P(R_t \mid \overline{H}_t)$$
$$\cdot P(\overline{B}_{Ut} \mid T_2 \ V_{23} \overline{X}_{It}) \cdot P(\overline{H}_t \mid \overline{X}_{It} T_2 V_{23}) \cdot P(V_{23} \mid T_2 \overline{X}_{It}) \cdot P(T_2 \mid \overline{X}_{It}) \cdot P(\overline{X}_{It})$$

$$(8.7)$$

Here:

$\overline{X}_\tau$: damage present at time τ;

H_τ : damage detected at τ;

R_τ : repair completed at t;

$\overline{B}_{Ut}$: Ultimate integrity lost at t

$\overline{H}_t$: Damage not detected at t;

$\overline{X}_{It}$: maintenance defect introduced at t

T_2 : defect type

V_{23} : maximum damage extent for type 2.

Example 8.6 This example provides an assessment of the order of magnitude of the contribution to the probability of an unsafe state due to maintenance defects in repair, and Equation (8.7) is used:

$$\Delta P(\overline{S}_T) = 0.3 \cdot 10^{-2} \cdot 0.5 \cdot 0.0014 \cdot 0.1 \cdot 0.10 \cdot 0.3 \cdot 10^{-6} = \text{small} \leqslant 10^{-13}$$

The limiting factor is the limit integrity implied by the required ultimate capability that must be restored by repair.

Missed large damage is for example a local defect in the size range D_5, which leads to load-path failure and, if not detected soon, will lead to PSE failure. The contribution to the probability of an unsafe state is described in the next example.

Example 8.7 The probability contribution is:

$$\Delta P(\bar{S}_T) = P(\bar{X}_\tau D_{5\tau}\bar{X}_I T_3 \bar{Y}_\tau R_{D\tau}\overline{HA}) = P(D_{5t} \mid \bar{X}_\tau) \cdot P(\bar{X}_\tau) \cdot P(T_3 \mid \bar{X}_I) \cdot P(\bar{X}_{1t})$$
$$\cdot P(\bar{Y}_t) \cdot P(R_{Dt}) \cdot P(\bar{H}) \cdot P(A_T)$$

This has a typical order of magnitude result:

$$\Delta P(\bar{S}_T) = 10^{-3} \cdot 10^{-2} \cdot 0.1 \cdot 10^{-6} \cdot 0.5 \cdot 1 = 0.5 \cdot 10^{-12}$$

which has a very low value, but if the defect is not caused by a true accident, but by a deliberate effort to suppress information, it could be much less, between 10^{-5} and 10^{-6}, which reinforces the importance of emphasizing the need for reporting.

The last two effects of defects due to maintenance result in mechanical damage, which for damage in the range of $\bar{D}_U$ causes loss of ultimate integrity, and for the size range D_5, loss of limit integrity. For the first case the contribution is

$$\Delta P(\bar{S}_T) = P(T_1 V_{12}\bar{X}_I \bar{B}_U) = P(\bar{B}_U \mid V_{12}T_1\bar{X}_I) \cdot P(V_{12} \mid T_1\bar{X}_I \cdot P(T_1 \mid \bar{X}_I) \cdot P(\bar{X}_I)$$

and numerical assessment

$$\Delta P(\bar{S}_T) = 0.2 \cdot 10^{-1} \cdot 10^{-1} \cdot 10^{-6} = 2 \cdot 10^{-9}$$

and for the limit range much less. The variety of damage sizes possible results in many sizes and a conservative, total effect in the order of magnitude of 10^{-8}.

8.3.1.3 Defects caused by operation

Effects of defects resulting from operation are:

1. exceeded limit load;
2. effects of extreme environment (e.g. hailstone encounters in flight);
3. large undetected or unreported mechanical damage present.

The first two defects caused by operation are temporary in nature. The first deals with the situation when external, limit load is exceeded and local structures are overloaded. The next example illustrates order of magnitude.

Example 8.8 We assume normal distributions for strength and loads and assume 20% overload, in which case the conditional probability of failure is

$$P(\overline{A} \mid T_1 V_{13} \overline{X}_O) = 10^{-5}$$

$$(8.8)$$

yielding

$$P(\overline{A}T_1 V_{13}\overline{X}_O B_{FS} R_D) = 10^{-5} \cdot P(V_{13} \mid T_1\overline{X}_O) \cdot P(T_1 \mid \overline{X}_O) \cdot P(\overline{X}_O)$$
$$\cdot P(R_D \mid B_{FS}) \cdot P(B_{FS})$$

where

B_{FS} : remaining structure has fail-safe integrity
R_D : load redistribution is successful

The contribution to the probability of an unsafe state is

$$\Delta P(\overline{S}_T) = 10^{-5} \cdot 0.3 \cdot 10^{-1} \cdot 10^{-3} \cdot 0.5 \cdot 0.9 \approx 10^{-10}$$

leading to a non-critical value probability.

8.3.1.4 Defects caused by requirements

Effects caused by defects in requirements are:

1. missing regulations
2. nebulous regulations
3. incorrect regulations.

Present transport-category regulations are the result of a philosophy that states:

Many existing 'metal' regulations can be used for composite material and structures.

A closer scrutiny reveals that this often is not an operative premise, and that its application often leads to either unsafe or conservative structures. Ultimate and limit allowable practices often lead to mean values for allowable and very conservative values, for example OH-compression and hot values. If we take an extreme, ultimate allowable value as the probability of violating ultimate integrity, we have the following combined event:

$$P(\overline{U}_U) = P(\overline{B}_U S_D S_E S_S) = S(\overline{B}_U \mid S_D S_E S_S) \cdot P(S_D) \cdot P(S_E) \cdot P(S_S)$$

Here:

the first factor is the probability that the design value is less than the applied internal loads;
The second factor is the probability of the prescribed state of damage;

The third factor is the probability of the prescribed environment, e.g. max temperature; The fourth factor is the probability of the prescribed, e.g. local 'stress' concentration due to existing or future fasteners.

The next example includes source of the scenarios used for ultimate allowable values.

Example 8.9 The use of 'hot, wet (late in service), open-hole allowable values' will result in probabilities such as:

$$P(\bar{U}_U) = 0.10 \cdot 0.99 \cdot 10^{-3} \cdot 10^{-2} \approx 10^{-6}$$

Here we have a B-value; no damage; maximum temperature; and open hole internal load distribution.

For panel (external) buckling:

$$P(\bar{U}_U) = 0.5 \cdot 0.99 \cdot 0.10 \cdot 0.9 = 0.044$$

Here we have mean value allowable; no damage; *RT*; no 'stress' concentration. Why are B-values, damage, $-65°F$ (if more critical), and evenly distributed loads used:

$$P(\bar{U}_U) = 0.1 \cdot 1 \cdot 0.7 \cdot 1 = 0.07$$

This example shows that, in composite structural design, there is a need to be subject to national design criteria.

The situation requires a regulated situation for composites, and interpreting this as a defect in the requirements process, we find that

$$\Delta P(\bar{S}_T) = P(\bar{U}V_{13}T_1\bar{X}_R) = P(\bar{U} \mid V_{13}T_1\bar{X}_R) \cdot P(V_{13} \mid T_1\bar{X}_R)$$
$$\cdot P(T_1 \mid \bar{X}_R) \cdot P(\bar{X}_R)$$

Here the numerical illustration yields:

$$\Delta P(\bar{S}_T) = 0.07 \cdot 0.3 \cdot 0.1 \cdot 10^{-6} \approx 2 \cdot 10^{-9}$$

With several analogous situations, in practice, the total contributions become $\approx 10^{-8}$.

Nebulous regulations include the enhancements provided for residual strength in the form of an AC for damage tolerance that has resulted in mean value residual strength often becoming the practice for metallic structures, and adopted for composites, even though damaged structures quite often have become critical for the probability of unsafe states and require B-values for safe designs.

The process for requirements development must be made safer by reducing the number of defects produced in:

- the process for new regulation;
- the selection of participants (design experience and education).

8.4 PURPOSE AND APPROACH TO STRUCTURAL INTEGRITY AND PROCESS QUALITY

The numerical illustrations in most of the previous chapters, and particularly those in the current chapter, have been structured to identify liabilities and assets in setting structural safety objectives and process qualities.

Damage tolerance is a very important part of composite structural safety. Limit integrity often dominates the identification of what is critical for design. Damage resistance, damage growth rates and damage tolerance interact to set requirements for limit damage sizes, low probability of large damage and high probability of detection, and they therefore influence allowable values of residual strength design data.

Ultimate integrity protects fail-safe integrity, and therefore is essential for protection from manufacturing and maintenance defects, which often come with damage that is not easily detectable. When detection does not yield adequate protection against load-path failure, fail-safety provides the protection necessary and limits the scale of the damage involved. The remaining structure will carry internal, ultimate loads and detection is a realistic possibility.

Safety protection, therefore, strives to keep both the probability of undetected loss of damage tolerance integrity and the undetected loss of ultimate integrity at low levels.

The process quality of all the elements of safety starts with a Six-Sigma quality with some drift, which makes improvements possible, rendering the characterization of new materials and processes iterative and rationally flexible to innovation, and allowing all the component parts of elements of safety to produce integrated requirements for all processes.

8.4.1 Analysis of numerical results

The focus of safety management is to establish, preserve and restore structural integrity and process quality, and to design, apply and update ad hoc processes for risk management, uncertainty mitigation and structural safety level control through analysis of service experience and to introduce them into the educational system.

The numerical results shown in Example 8.10 are particularly dependent on the practical and necessary attainment of safety levels for safe states and requirements for quality-driven defect frequency and reliability of monitoring and updating processes: the first item, through explicit, safety-based design constraints; the second item, through the use of improved Six-Sigma quality requirements; the third item through informed updating of safety levels.

The value of the 'Realistic' numerical values and the functions and processes that have been used in the previous scenarios and analyses include the following:

- The inspection of structures designed to have a low probability of an unsafe state will use inspection methods and frequency to find damage, before load-paths fail, in scheduled inspection, or it will be designed fail-safe when one load path fails and subject to walk-around inspection.
- Many defects associated with elements of safety are very hard to detect and also require fail-safe design.
- Process development must give the true relations between defects, effects, total extent and what is unacceptable overall for safety level goals.
- The frequencies of inflicted 'accidental' damage, the occurrence of process defects from all the elements of safety and the probability of detection of different size enter into the final value of the probability of an unsafe state.

Example 8.10   This example is based on FAA-supported recommendations from Vice-President Gore's Commission on Safety of Aviation published in the 1990s:

For an airplane: .... one unsafe flight in 10^5: .... 10^{-5}
For structure, 10%: .... 10^{-6}
Improvement by 2005, 10%: .... 10^{-7}

Per PSE, of 40:

Per DDP (average), 25: ....10^{-3} .... 10^{-10}
Uncertainty: ....10^{-11}

This is obviously a description of a philosophy, subject to the decisions of the original equipment manufactures, the FAA and the International Harmonization Community.

8.4.2   Unsafe state – large damage

The related design criterion is based on the philosophy of minimizing the probability of an undetected loss integrity' (both damage tolerance and fail-safe)

$$P(\bar{S}_T) = P(\bar{H}_\tau \bar{U}_T \bar{H}_T) = P(\bar{H}_\tau) \cdot P(\bar{B}_L \mid \bar{H}_T \bar{X}_T D_{5T}) \cdot P(\bar{H}_T \mid D_{5T} \bar{X}_T)$$
$$\cdot P(D_{5T} \mid \bar{X}_T) \cdot P(\bar{X}_T) \tag{8.9}$$

Equation (8.9) describes an event that covers two scheduled inspections, τ and T.
The events are:

$\overline{H}_\tau$: undetected damage at τ, The marginal distribution is used as DT – quality for metal structure

$\overline{B}_{LT}$: residual strength, $RS < LLR$ (limit load requirement)

$\overline{H}_T$: not detected at T

$\overline{X}_T$: damage present at T

D_5 : damage size in limit region.

Equation (8.9) will now be used in the following example.

Example 8.11 A typical situation is now assessed. A set of required values are used:

$$P(\overline{S}_T) = 10^{-2} \cdot 10^{-1} \cdot 10^{-3} \cdot 10^{-3} \cdot 10^{-2} = 10^{-11}$$

Here the first factor is the marginal probability of not detecting damage. The second represents the B-value residual strength. The third is the overall probability of not detecting damage in size range D_5. The fourth is the probability of having a limit size damage range. The last value represents the probability that damage is present. The consequence of satisfying the quoted requirements results in some very stringent requirements.

The remaining consideration, for this case, is how long the situation with undetected lost integrity can stay in a safe range with a 'reasonable' probability. We will now assume that normal distributions are involved and ask: 'How long can the structure survive without failure?'

Probability of failure after n flights

$$P(\overline{A}_n) \approx P(\overline{A}_n \overline{S}_T) = P(\overline{A}_n \mid \overline{S}_T) \cdot P(\overline{S}_T) = P_{SUn} \cdot P(\overline{S}_T)$$

Here P_{SUn} is the probability of surviving n flights, P_{SU}^n

The probability of failure for one flight with lost integrity is based on B-value residual strength and 2σ quality loads. The normal assumption leads to a normal distribution that predicts failure when the variable S–L (strength–load) < 0:

$$p_{SU}^1 = 1 - p_f^1 = 1 - \Phi\left(\frac{0.45}{0.13}\right) = 1 - 0.00023 = 0.99977$$

Survival for n flights in an unsafe state can be shown as:

n	p_{SU}^n	Probability of failure at $\leq n$ flights
1	0.99977	–
1000	0.79	0.21
3000	0.50	0.50
6000	0.25	0.75
12000	0.06	0.94

The situation is 'marginal' and the choice of improved inspection methods may improve the situation.

There is a not uncommon 'practice', in the composites world, of using 'mean quality allowable data' in design. The consequence is:

n	p_{SU}^n	Probability of failure at $\leq n$ flights
1	0.9938	–
1000	0.002	0.998
3000	10^{-8}	~1

Obviously this is an unsafe practice.

8.4.3   *Process defects – influence from all defects originating from elements of safety*

The effects of defects can most often be categorized as 'loss of integrity, no detection, rescue by fail-safe design'. A special class of scenarios associates n flights with single load-path failure, preserved fail-safe integrity, and continuing N flights till final detection.

The example below involves a manufacturing defect which violates the residual strength objective. The equation describing this situation includes failure of load-path at time t.

Example 8.12   The equation below is directly applicable to 'poor detection'.

$$P(\bar{S}_t) = P(\bar{U}_o \bar{X}_M V_{ij} T_i \bar{Y}_t R_t U_t \bar{A}_n) = P(\bar{U}_o \mid \bar{X}_M T_i V_{ij}) \cdot P(V_{ij} \mid \bar{X}_M T_i)$$
$$\cdot P(T_i \mid \bar{X}_M) \cdot P(\bar{X}_M) \cdot P(\bar{A}_n \mid \bar{Y}_t R_t U_t) \cdot P(U_t \mid \bar{Y}_t R_t) \cdot P(R_t \mid \bar{Y}_{t|}) \cdot P(\bar{Y}_t)$$

$$(8.10)$$

Here the factors on the right-hand side are:

Probability of lost ultimate integrity due to manufacturing defect
Probability of degree of defect
Probability of type of defect
Probability of process failure
Probability of load-path failure in n flights
Probability of preserved fail-safe integrity after load-path failure
Probability of load redistribution
Probability of failure of a fail-safe load-path.

This situation is dominated by a manufacturing defect (loss of structural properties) that compromises ultimate integrity as described by the first four factors and that is associated with limited detectibility which is often likely to lead to load-path failure. If the fail-safe design has the proper integrity, the internal loads will become redistributed, external limit load capability will be re-established, and a better detectibility will be created. The numerical evaluation of Equation (8.10) raises a number of issues. The process defect must be localized to the area of one DDP, for which the equation will produce the following results:

$$P(\bar{S}_T) = 10^{-1} \cdot 10^{-1} \cdot 10^{-2} \cdot 10^{-9} \cdot 1 = 10^{-13}$$

This result could be considered acceptable, but does not allow for any 'drift' in the Six-Sigma quality in the process.

If, however, the defect is localized, we would arrive at the event with the following prognoses. If a load-path has failed, the two adjacent ones will have retained their ultimate integrity with a local region at ultimate internal load and normal distributions for strength and internal loads; should the failed load-path not be detected, the probability of an unsafe state increases as:

Number of flights (N)	Probability of failure before the N'th flight (factor on state just after single load-path failure)
1000	0.24
3000	0.57
6000	0.88

8.4.4 Structural safety

Structural safety threatened by mechanical (accidental) damage is measured in terms of probability of an unsafe state. It is afforded protection by reason of the low probability of an unsafe state, by detection and by walk-around inspections whenever possible and, as a backup, by fail-safe design which, combined with damage resistant design, protects both damage tolerance and fail-safety.

Structural safety is also threatened by process defects (often difficult to detect), measured by probability of loss of integrity, and protected by fail-safe design, which also improves detectibility.

The numerical investigation indicates that the process quality requirements have to satisfy some version of Six-Sigma, reinforced for the design process by the required test programs and the certification compliance demonstrations.

Process defects of a widespread nature, such as process defects in manufacturing, present a difficult situation, because they often threaten the fail-safe integrity, leaving quality assurance and quality control as the protective features. For example, curing and co-curing are processes for which statistical process control seems to be a very promising development, especially in consideration of the fact that the safety requirement formulation must include the relations between the nature of defects and the effects on structural performance.

8.4.5 Safety control

This chapter has addressed the 'weapons of safety' available; the following is the list of what is effective and required to meet the safety requirements as illustrated by the numerical investigations in previous chapters:

- Damage tolerance critical structure;
- Minimum probability of an unsafe state (undetected loss of damage tolerance integrity);
- Significant time of survival after loss of damage tolerance integrity;
- Damage resistance minimum capability;
- Maximum damage growth rates to be established;
- B-value residual strength is the consequence of the required safety level;
- Effective inspection for D_5 size region $P(H/D_5) = 10^{-3}$ and walk-around (whenever) make B-values possible;
- extra attention to repairs;
- high-quality processes (localized defects) – Six-Sigma processes recommended;
- fail-safe design (damage tolerance criticality preserves fail-safe integrity with a margin as high as 20%)
- Fail-safe improves detection before failure.
- Widespread defects in, for example, curing and co-curing, causing loss of fail-safe integrity, often being hard to detect. The remedy is a process control and quality control process that detects defects that cause a reduction in structural properties of greater than 20%.

All these point are integrated as part of design criteria work, which includes requirements for damage characterization and for probability of detection, for damage resistance, for fail-safety of loaded structure and for survival of structure with loss of limit integrity or failed single load-path.

8.5 CONCLUSIONS

The critical mode of composite structural design is damage tolerance:

- It should be founded on damage resistance and supported by a detailed characterization of maximum damage growth rates.
- Fail-safe design is the preferred fall-back when *localized* process defects are produced by the other elements of safety.
- Ultimate integrity should allow a margin of up to 20% to deal with extreme structural property defects.

Widespread process defects must pass through process and quality controls that are good enough to prevent these defects from entering service.

Chapter 9
Structural Service Monitoring

The worlds of composite and hybrid structures are in a state of constant innovation. New materials, processes, structural concepts and metallic interfaces are steadily appearing. New flight vehicles pass in review, and service experience and empirical design databases are accumulating slowly. Innovation is the operative word, and baseline design data are derived from what is produced in the development, concept formulation and design phases, making the foundation uncertain, incomplete and in need of change. In the commercial airplane world prototypes are not regularly used, instead there is an intense need to monitor service performance to reduce risk, mitigate uncertainty and maintain safety levels. Between full-scale structural testing and flight tests, much is achieved toward good safety levels. Because of the lack of service experience for the recent newcomers in the 'composites world', there is much service monitoring needed to avoid surprises, especially for the transport category, but not limited to that category.

The monitoring process must specifically collect service data of a random nature to perpetually update probabilities of damage detection, to determine frequency of occurrence and to update and improve the understanding of the probabilistic nature of the service situations.

The monitoring process is seen as focused on each principal structural element (PSE) – that is, considering a model compatible with the distribution of detail design points (DDPs), so that damage, defects, detection probabilities, internal loads, and structural geometry all have the same reference points.

The data that take priority in the monitoring process deal with the probability of different sizes of damage and their detectibility. The regions for different size ranges are described in Chapter 8, and the data collected are for the DDPs of the PSE in question. The data describe probability of damage present and the probability of size range, given that damage is present; the range of sizes covers everything except ultimate damage size.

The focus of the process is to analyze the number in each damage size region and report cumulative information.

$$P(H\bar{D}_U D_i \mid \bar{X}_U) = P(H \mid \bar{D}_U D_i \bar{X}_U) \cdot P(\bar{D}_U D_i \mid \bar{X}_U) \tag{9.1}$$

Here:

The combined event represents probability of detection given damage in D_i, considering

$$\bar{D}_U = D_3 \cup D_4 \cup D_B \cup D_5 \cup D_{FS} \cup D_6$$

Here:

D_B : a 'buffer region' for growth

D_{FS} : the region describing fail-safe, load-path incapacitated and internal loads redistributed

D_6 : the region of excessive damage size.

9.1 MONITORING PROCESS EXAMPLES

The process objectives and requirements are a combination of what is desirable for different kinds of material systems and processes; this section focuses on consequences which are realistic in the composites world. We start by illustrating a few situations that 'push the envelope'.

For the case when the cumulative count is increased by two sets of inspection results in region D_5, we have the following illustration (Example 9.1).

Example 9.1 We assume that the PSE has 50 DDPs and that two counts are added in D_5, making the total two during the previous inspection periods. First we assume that the damage count is increased by two soon after the last inspection, leading to the following equation for the probability of two counts in D_5:

$$P(D_5 \text{ contains } 2 \text{ counts}) = \frac{2}{50} \cdot \frac{1}{49} = 0.0004 \Rightarrow P(\bar{A_1}) = 0.0005$$

We assume two DDP's with unsafe states and calculate the probability of failure for a different number of flights:

n	p_{su}	Probability of failing in less than or equal to n flights
1	0.99954	–
100	0.955	0.045
1000	0.63	0.37
3000	0.25	0.75

Example 9.2 We assume that there are warning signs delivered when analyzing inspection results using 'prior assumptions'. For example, the probability of an unsafe state when damage in D_5 is detected is a valuable indicator.

Assume that there are two counts of damage in size range D_5 and only one detected. The probability of unsafe states is then

$$P(\overline{S}) = \tbinom{50}{2} \cdot 10^{-2} \cdot 10^{-6} \cdot \tbinom{2}{1} \cdot 0.999 \cdot 10^{-3} \approx 2.45 \cdot 10^{-8}$$

Or if we look at the case of three counts of damage in size range D_5 and only one is detected, we find

$$P(\overline{S}) = \tbinom{50}{3} \cdot 10^{-2} \cdot 10^{-9} \cdot \tbinom{3}{1} \cdot 0.999 \cdot 10^{-6} \approx 0.6 \cdot 10^{-12}$$

Discovering damage in D_5 should encourage caution because the risk for an unsafe state at the beginning of an inspection period is marginally larger for some events, and should result in repeating the local inspection. This is a good case for keeping accumulative inspection records for each PSE.

The probability of damage being present, $P(\overline{X})$, for size range $\overline{D}_U$, can also be a good tool for risk identification and uncertainty reduction. If combined with comparisons between accumulated data for PSEs, it would reveal variations between different PSEs and what actions to take.

The total set of detectable damage sizes, $\overline{D}_U$, also contains the regions D_6, excessive damage and D_{FS}, and failed load-path in fail-safe regions, for which the accumulative records would reveal both local variations and something being wrong.

The records discussed in Example 9.1, especially data changing in time, would be rendered a great deal more certain in time and more understandable, when, for example, 'drift' in process quality control is present. These data would also be a treasure trove of service experience, input to empirical design upgrades and education (e.g. cooperative know-how. Finally, it would be a very effective source for Bayesian updating.

Example 9.3   Here is given an illustration of a Bayesian update of probabilities of a group of all damage sizes. Data from the first inspection and accumulative data from the first three inspections relating to one PSE are used.

i	x_i	Prior	Likelihood	Prod.	Posterior	Prod.	Posterior
–	–	–	1	1	1	first 3	first 3
1	3	0.1	0.1	0.01	0.9874	0.036	0.90
2	4	10^{-2}	0.01	10^{-4}	0.0097	0.0033	0.0825
3	B	$5 \cdot 10^{-3}$	0.005	$25 \cdot 10^{-6}$	0.0025	0.00018	0.0045
4	5	10^{-3}	0.003	$3 \cdot 10^{-6}$	0.0006	0.00009	0.0022
5	6	10^{-4}	0	0	0	0.000003	10^{-4}

The table is arranged with the second column containing the indices for the damage size region, the third column the prior probabilities, the sixth column the posterior probabilities derived from the first inspection and the eighth column the posterior probabilities derived from the results of the first three inspections.

The purpose of this example is to emphasize how uncertainty can be decreased with time by using service data to update prior probability distributions.

9.2 PRIORITY TARGETS FOR MONITORING

Data particularly important to monitor, report and record are those for the purpose of updating data used in the design process and for diagnostic use to identify need for 'better' data and regions with potential problems. Returning to the definition of the structural elements of safety that we have discussed and accounted for, we have the following account:

Design: the following sources of safety threats are examples of identified defects in the *design process for composites*:

- critical conditions not considered and not covered in structural full-scale testing (e.g. low atmospheric temperatures at high altitude);
- serious combinations of damage states, environmental states and local 'stress-rises' not identified in FE analyses, and not included in the compliance testing.

The following sources of safety threats are examples of identified defects in the *design execution for composites*:

- not identifying a realistic range of damage sizes and probabilities of occurrences, which are consequently not included in the testing;
- not doing fail-safe testing (without load and under load), in the compliance demonstration for certification.

Manufacturing: the following sources of safety threats are examples of identified defects in the *manufacturing processes for composites*:

- not identifying the realistic range of defects that reduce structural properties and threaten structural integrity, or producing the test data for the accept/reject criteria;
- not identifying how widespread process defects are occurring and what the probability of detection is.

The following sources of safety threats are examples of potential defects in the execution as selected in the *manufacturing execution of composite processes*:

- Widespread process failure leading to loss of fail-safe integrity in a region, if not detected in quality control (could be acceptable up to the level of property reduction protected by the ultimate margin of safety);

- Load increase (excessive fastener loads or bond-line internal load variations), due to poor fastener installations, bond-line thickness variations, lost torque and excessive clamp-up.

Maintenance: the following sources of Safety threats are examples of potential defects in the *maintenance process for composite structure*:
- increase in internal loads due to inadequate scheduled maintenance (e.g. lubrication);
- poor fastener installation or bonding.

The following sources of safety threats are examples of *defects in execution of maintenance processes for composite structure*:
- damage inflicted in structure during execution of the maintenance repair process during removal procedures in damaged areas;
- damage caused in structure by walking on sensitive areas as part of all maintenance processes.

Operation: the following events are examples of *defects in operational processes* that could threaten the integrity of a composite structure:
- exceeding internal limit loads due to mistakes in the management of external loads;
- not avoiding extreme flight environments (e.g. hailstorms).

The following sources of safety threats are examples of defects in the *execution of operational procedures* (processes):
- not avoiding severe turbulence and inflicting structural damage;
- not following regulated procedures in flying and training.

Requirement formulation: the following sources of safety threats are examples of defects in requirement processes with emphasis on regulations:
- specific regulations for composite structure – needed to focus on the specifics of composites and their sensitivities;
- regulations for composites' allowable values and design data.

The following sources of safety threats are examples of the defective use of existing regulations relating to metals:
- inadequate use of B-values for fail-safe structure;
- new definitions of limit load needed to design composite structure safely.

9.3   STRUCTURAL DEFECTS

The Elements of Safety all produce a wide variety of defects that potentially either must be avoided, found or included in the Design Requirements. The numerical illustrations in a large number of the previous chapters have demonstrated that all the defects can be lumped under one or several of the following headings.

The elements of safety potentially produce defects in processes or implementation, constituting threats to the following design data. The headings are:

- reduced structural properties;
- spurious, internal, design loads;
- misleading, missing or nebulous design requirements;
- mechanical damage.

The safety threats associated with these four fields are either controlled by process quality or containments of the probability of unsafe States, or both. Each entry in these headings is driven by one or more elements of safety. For example, reduced structural properties can be the results of manufacturing defects or maintenance problems.

The requirements are based on equal burden being borne by all the elements of safety in relation to the probability of an unsafe state.

9.3.1 *Reduced structural properties*

The contribution to structural property reductions made by manufacturing process defects can be kept under control by effecting Six-Sigma process quality, assuming that detrimental effects are relegated to outside the 6σ zone, that widespread reductions are discovered by the quality control function and that the PSE count fits the within the regime used in the objective. The model for design is based on DDPs, and if each DDP is independent then the safe state for a PSE, S_P, can be written in terms of safe states at the DDPs, S_{Di}

$$P(S_P) = P(S_{D1} \cdot S_{D2} \cdot \cdots S_{Dn}) = P(S_{D1})P(S_{D2}) \cdot \cdots \cdot P(S_{Dn})$$
$$P(\bar{S}_P) \approx P(\bar{S}_{D1}) + \cdots + P(\bar{S}_{Dn}) \tag{9.2}$$

while widespread defects of a significant part of a PSE increase the probability of failure. Widespread defects in the PSEs, like the ones mentioned, must be found before they are incorporated into the structure.

The next example illustrates how a Six-Sigma quality satisfies the safety requirements in relation to reduced strength.

Example 9.4   We deal here with the joint event describing the probability of a safety threat due to a process failure in manufacturing in the presence of Six-Sigma enforced quality. The requirement used for a DDP is, when Equation (9.2) applies, $<2.5 \cdot 10^{-11}$, (40 DDPs); when Equation (9.3) applies, the requirement is less stringent for DDPs.

The probability of a manufacturing/material process failure and the quality is governed by a Six-Sigma requirement:

$$P(\bar{S}) = P(\bar{X}_M T_1 V_{13} \bar{M}_{PC} \bar{M}_{QC}) = P(\bar{M}_{QC} \mid \bar{M}_{PC} V_{13} T_1 \bar{X}_M) \cdot P(\bar{M}_{PC} \mid V_{13} T_1 \bar{X}_M)$$
$$\cdot P(V_{13} \mid T_1 \bar{X}_M) \cdot P(T_1 \mid \bar{X}_M) \cdot P(\bar{X}_M) \qquad (9.3)$$

Here the following events describe a loss of strength, and are:

$$\bar{X}_M \quad : \quad \text{manufacturing process defect}$$
$$T_1 \quad : \quad \text{processing failure}$$
$$V_1 \quad : \quad \text{defect outside Six-Sigma zone}$$
$$\bar{M}_{PC} \quad : \quad \text{missed in process control}$$
$$\bar{M}_{QC} \quad : \quad \text{missed in quality control.}$$

The numerical illustration in the order of the above equation:

$$P(\bar{S}) = 10^{-1} \cdot 10^{-1} \cdot 10^{-1} \cdot 10^{-2} \cdot 10^{-8} = 10^{-13} \le 10^{-11}$$

which would allow for some drift.

This allows for another source of reduced properties, e.g. maintenance, where failure of processing of bonded repairs would produce a contribution to the probability of an unsafe state of the same order of magnitude as above.

Even though a failure in repair processing can increase the probability of an unsafe state by producing strength reduction in bond-lines, or reduced strength due to a local increase in fastener loads, and the requirement for a safe state is satisfied, one would expect that special attention is given to repair inspections considering the risk for propagation of damage in concealed locations.

The arguments for compensating for strength reductions are valid against stiffness reductions because of the effects on buckling. This also makes intrusions into the margins on dynamic instabilities and should be investigated.

Damage resistance, **damage growth rates** and **damage tolerance** form a group of structural properties that influence residual strength and are especially important to structure that is design-critical in its damaged state. It is often true that all three are affected at the same time. Reduced damage resistance causes larger initial accidental damage sizes, increased damage growth rates cause larger damage and reduced damage tolerance causes earlier failure.

The damage shown in Figure 9.1 starts out larger than D_4, and grows faster. Only one detection opportunity arises before the single load-path failure and must be ensured by fail-safe design.

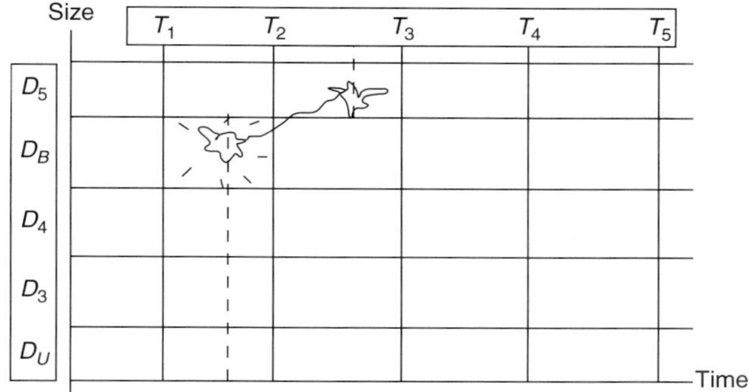

Figure 9.1 Reduced damage resistance.

The probability of an unsafe state is

$$
\begin{aligned}
P(\overline{S}_t) &= P(D_{Bi}\overline{X}_{M1}T_1V_{13}D_{5t}\overline{X}_{M2}T_2V_{23}\overline{H}_{T2}\overline{U}_t\overline{A}_n) \\
&= P(D_{Bi} \mid \overline{X}_{M1}T_1V_{13}) \cdot P(D_{5t} \mid D_{Bi}\overline{X}_{M2}\,T_2V_{23}) \cdot P(\overline{H}_{T2} \mid \overline{U}_t) \\
&\quad \cdot P(\overline{A}_n \mid \overline{U}_t) \cdot P(V_{13} \mid T_1\overline{X}_{M1}) \cdot P(T_1 \mid \overline{X}_{M1}) \cdot P(\overline{X}_{M1}) \\
&\quad \cdot P(V_{23} \mid T_2\overline{X}_{M2}) \cdot P(T_2 \mid \overline{X}_{M2}) \cdot P(\overline{X}_{M2})
\end{aligned}
\tag{9.4}
$$

Here we have the following participating events:

D_{B1}	:	initial damage is in region D_B
$\overline{X}_{M1}$	:	manufacturing defect present in damage resistance
T_1	:	defect type is decreased damage resistance
V_{13}	:	severe reduction
D_{5t}	:	damage in region 5 at time t
$\overline{X}_{M2}$	:	defect in damage growth rate
$\overline{H}_{T2}$	:	damage not detected at T_2
$\overline{U}_t$	:	loss of limit integrity at t
$\overline{A}_n$	:	failure of n flights after time t.

Figure 9.1 describes a scenario where the defect is limited to the load-path in question, or it is demonstrated that reduced damage resistance does not compromise the fail-safe, internal load redistribution required for fail-safe integrity. The order of magnitude expected from Equation (9.4) is evaluated in the next example.

Example 9.5 The objective of this example is to demonstrate what the 'requirements' of the probabilities shown on the right-hand side of Equation (9.5) are, for the case when only damage resistance and damage growth rates are affected.

$$
P(\overline{S}_t) = 10^{-2} \cdot 0.25 \cdot 10^{-3} \cdot p_s^n \cdot (1 \cdot 0.1 + 0.3 \cdot 1) \cdot 10^{-6} = p_s^n \cdot 10^{-11}
$$

Here p_S^n, = survival for n flights, which does not allow more than 1000 flights of failure with a probability of 40%, or possibly less, has promise of damage resistance and growth rates are not both linked to the same processing defect.

This again illustrates that the safety analyses and designs of materials, processes and structural concept must be done on a case-by-case basis, and is an important part of materials and processing selection criteria.

Example 9.6   We assume that the defect, for the same damage size, is reduced to 80% of uncompromised value and that both strength and internal loads distributions are normal. The probability of failure for a random flight becomes 0.035, which leads to a failure in 100 flights or less of ~ 0.97.

The probability of a processing failure that reduces the residual strength to 80% of acceptable values, the presence of damage that belongs in size range D_5 and failure within 100 flights of loss of limit integrity is

$$P(\overline{A}_{100}) = P(\overline{X}_M T_3 V_{33} \overline{X}_t D_{5t} \overline{p}\,_f^{100}) = P(V_{33} \mid T_3 \overline{X}_M) \cdot P(T_3 \mid \overline{X}_M)$$
$$\cdot P(\overline{X}_M) \cdot P(D_5 \mid \overline{X}_t) \cdot P(\overline{X}_t) \cdot 0.97 \qquad (9.5)$$

The numerical illustration becomes

$$P(\overline{A}_{100}) = 10^{-2} \cdot 0.1 \cdot 10^{-6} \cdot 10^{-3} \cdot 10^{-2} \cdot 0.97 \approx 10^{-14}$$

This result captures the importance of using the right scenario in evaluating the structural safety threats associated with manufacturing defects, and how they are interacting. It also emphasizes the importance of characterizing the details about what defects cause which property changes.

9.3.2   *Spurious internal design loads*

Situations causing spurious internal loads in the processes associated with the elements of safety often also produce results that are difficult to detect. Three situations of a different nature arise:

1.   exceeding external limit loads due to operational actions;
2.   geometric variations (causing local concentration);
3.   assembly defects causing self-equilibrating internal loads that are additive.

9.3.2.1   **External limit load exceeded due to operational defect**

This is a rare event that violates proper operating procedure and would have to coincide with lost limit integrity, but could affect many PSEs because the specific load case might be critical in many locations. The most natural counter-measure is to design an operational process that is based on, for example, Six-Sigma.

The probability of an unsafe state in a PSE can be written as

$$P(\overline{S}_{ij}) = P(\overline{X}_O T_3 V_{31} \overline{X}_t D_{5t}) = P(V_{31} \mid T_3 \overline{X}_O) \cdot P(T_3 \mid \overline{X}_O) \cdot P(\overline{X}_O)$$
$$\cdot P(D_{5t} \mid \overline{X}_t) \cdot P(\overline{X}_t) \tag{9.6}$$

Here the following events are involved:

$\overline{S}_{ij}$: unsafe state at PSE i and DDP j
$\overline{X}_O$: operational defect
T_3 : external limit load exceeded
V_{31} : event lasts until equilibrium is reached and persists
$\overline{X}_t$: mechanical damage is present
D_{5t} : damage size is in region D_5.

Example 9.7 The nature of the event in Equation (9.6) is such that the specific load case in question can be critical at several PSEs and at each one of these at several DDPs, making the case-by-case situation complicated.

We assumes that n_p are affected and that some of these have mechanical damage in the D_5 size range, n_D.

A numerical illustration of a DDP can be given with an evaluation of Equation (9.6), one result being:

$$P(\overline{S}_{ij}) = 10^{-2} \cdot 10^{-1} \cdot 10^{-9} \cdot 10^{-3} \cdot 10^{-2} = 10^{-17}$$

Even if 10 PSEs are involved and each one has 3 damage sizes in D_5, we have

$$P(\overline{S}) = 3 \cdot 10^{-15}$$

We could limit the process quality to 4.5 Sigma with a result of 10^{-12}. This requirement depends on the number of defects and the number of affected PSEs, and must be part of the monitoring process as the probability that there are three D_5 sizes in a PSE, given that damage is present, $P(\exists\ 3\ \text{defects in}\ D_5 \mid \overline{X})$, is:

$$\binom{N}{3} p_{D5}^3 \cdot \overline{p}_{D5}^{(N-3)} = 0.4 \cdot 10^{-5}$$

9.3.2.2 Increased internal loads due to defects in geometry

Geometric defects causing internal stress/strain concentrations can be caused by both manufacturing (installation, assembly, co-bonding and co-curing) and repair defects and must be countered by Six-Sigma process quality in manufacturing and repair, combined with Six-Sigma operation process quality that limits internal loads defects to $1.05 \cdot$ LLS. We then need to count on detection and fail-safe to guarantee safety levels, as the presence of a defect does not assure adequate survival. Those defects that involve self-equilibrating internal load systems must be dealt with in a similar fashion.

Operational mistakes such as exceeding limit load or failing to evade extreme environmental conditions are best avoided through training and education, discipline and reporting. The uncertainty relating to defects and random conditions, along with reliable reporting and accumulation of service experience are difficult challenges to meet, but it should always be borne in mind that the potential for disastrous results requires continuous vigilance, especially in education and training. The task of designing an n-defect situation is complicated.

9.4 CONCLUSIONS

Defects resulting from different processing failures associated with all elements of safety and the effect on structural performance characteristics vary from case-to-case and should be part of the characterization of materials and processing.

The most effective way to achieve safety levels is to implement Six-Sigma process quality with back-up detection (walk-around detection capability), and fail-safe capabilities in the local design requirements.

The demonstration of damage resistance compliance is a very important part of safety.

Chapter 10
Structural Integrity, Safety and Design

A central issue in the design of composite structures is damage tolerance. Composite structures are often in a critical state when 'large' damage is present, loss of damage tolerance integrity constituting an 'unsafe state' inevitably leading to failure if not detected and corrected. The distribution of 'operating' internal loads is an elusive quantity, such loads varying from location to location; many similarities with stress fatigue spectra can be seen and the best way to deal with this situation might be in terms of criteria.

Existing practices are not uniform, and future regulations need a baseline of structural safety. This could be approached through the following equation, which describes the probability of an unsafe state:

$$P(\bar{S}) = P(\bar{A}_n \bar{X}_t D_{5T} \bar{B}_{LT} \bar{H}_T) = P(\bar{A}_n \mid \bar{X}_t D_{5T} \bar{B}_{LT} \bar{H}_T) \cdot P(\bar{H}_T \mid \bar{X}_T D_{5T} \bar{B}_{LT})$$
$$\cdot P(\bar{B}_{LT} \mid D_{5T} \bar{X}_t) \cdot P(D_{5T} \mid \bar{X}_t\} \cdot P(\bar{X}_t) \tag{10.1}$$

$\bar{A}_n$: Failure after n flights
$\bar{X}_t$: Damage present since time t
$\bar{H}_T$: Damage not detected at time T
$\bar{B}_{LT}$: Residual strength less than requirement at time T
D_{5T} : Damage size in region D_5 at time T.

Numerical values with a realistic chance of satisfying the safety requirements yield

$$P(\bar{A}_n \bar{S}_T) = (1 - p_s^n) \cdot 10^{-8} \cdot P(\bar{B}_{LT} \mid D_{5T} \bar{X}_t)$$

There is not a consistent practice for residual strength allowable values. If we adopt, for background, the ultimate allowable practice we get the following equation:

$$P(\bar{A}_n \bar{S}_T) = (1 - p_s^n) \cdot 10^{-3} \cdot P(\bar{B}_{LT} \mid D_{5T} \bar{X}_t S_E S_s) \cdot P(D_{5T} \mid \bar{X}_t)$$
$$\cdot P(\bar{X}_t) \cdot P(S_E) \cdot P(S_S)$$
$$= (1 - p_s^n) \cdot 10^{-8} P(S_E) \cdot P(S_S) \cdot P(\bar{B}_{LT} \mid D_{5T} \bar{X}_t S_E S_s) \tag{10.2}$$

10.1 STRUCTURAL SAFETY AND CRITERIA

The first set of practices involves the choice of allowable definitions. We will compare the numerical results from three alternatives.

Example   10.1 We assume normally distributed strength and internal load variables; the first case involves B-values, and $C_V = 0.1$. The stress requirement is LLS.

for Strength we get

$$\mu_x = 1.15\text{LLS} \quad \text{and} \quad \text{loads yield } \mu_y = 0.7\text{LLS}$$

The probability of safety consequence is

$$\mu_g = (1.15 - 0.7)\text{LLS} = 0.45\text{LLS} \quad \sigma_g = (1.32 + 0.49)^{1/2}0.1\text{LLS} = 0.13\text{LLS}$$

and probability of failure is

$$\Phi(-0.45/0.13) = \Phi(-3.46) = 0.00023$$

yielding a probability of survival

$$p_s = 0.99977$$

The results for different numbers of flights are:

n	Probability of surviving n flights	Probability of failure in $\leq n$ flights
1	0.99977	0.00023
1000	0.79	0.21
3000	0.50	0.50
6000	0.25	0.75

Example 10.2 If we instead use mean value for allowable, the probability of safety consequences are

$$\mu_g = (1.0 - 0.7)\text{LLS} = 0.3\text{LLS} \quad \text{and} \quad \sigma_g = (1 + 0.49)^{1/2}0.1\text{LLS} = 0.12\text{LLS}$$

The probability of failure is:

$$\Phi(-30/0.12) = \Phi(-2.5) = 0.0054$$

And the probability of survival becomes: 0.9946

n	Probability of surviving n flights	Probability of failure in $\leq n$ flights
1	0.9946	0.0054
1000	0.0045	0.9955
3000	~ 0	~ 1

From this, one could conclude that mean allowable values are inadequate for the safety level deemed necessary, and the probability of an unsafe state is

$$0.5 \cdot 10^{-11} \text{ at } n = 1$$

and that the probability of surviving 1000 flights is 0.0045, which would make B-values a requirement for residual strength, if a fail-safe detail design were not implemented.

10.1.1 Damage growth and criteria

Where a buffer region, D_B, is used for control of aberrant growth into the limit size region, it is of interest to understand the safety situation. Figure 10.1 illustrates a situation when all regions are of equal length, the following example describing the situation.

Example 10.3 Again we assume normal distributions and $C_V = 0.1$. The strength and load parameters are

$$\mu_X = 1.29\text{LLS} \quad \text{and} \quad \mu_Y = 0.7\text{LLS}$$

yielding the parameters for probability of failure:

$$\mu_g = (1.29 - 0.70) \cdot \text{LLS} = 0.59\text{LLS} \quad \text{and} \quad \sigma_g = (1.66 + 0.49)^{1/2}0.1\text{LLS} = .15\text{LLS}$$

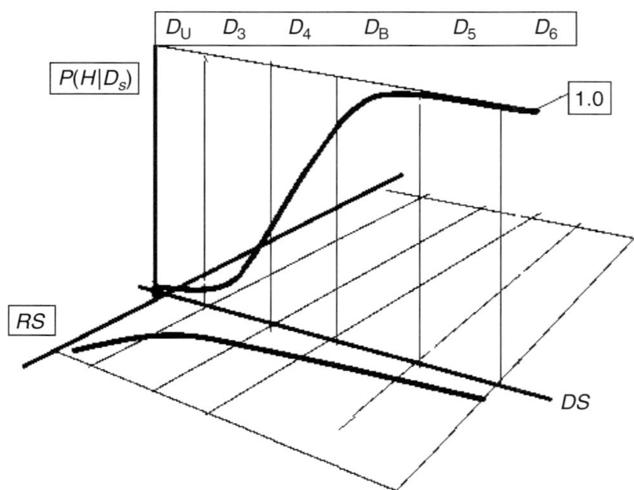

Figure 10.1 Illustration of residual strength and detection.

The probability of failure is: $\Phi(-0.59/0.15) = \Phi(-3.9) = 0.00005$, and the survival probability is

$$ps = 0.99995$$

The probabilities for different numbers of flights are shown in the table below. The table describes a range of possible inspection periods:

n flights	Probability of surviving n flights	Probability of failure in $\leq n$ flights
1000	0.95	0.05
3000	0.86	0.14
6000	0.74	0.26
9000	0.64	0.36
12000	0.55	0.45

The probability of not detecting damage in for example, four inspection periods (3000 flights per period) is between 10^{-9} and 10^{-8} (based on the value at the ends of the intervals D_4, D_B, D_5), which illustrates the importance of defining damage size regions and criteria wisely.

10.1.2 Mechanical damage and criteria

Design	:	exposing poor design details or concepts
Manufacturing	:	process failures (in bonding), tool drops, transport impacts etc.
Maintenance	:	stepping on fragile structure (in scheduled maintenance, in inspection, in repair), tool drops etc.
Operation	:	impact by ground vehicles, runway debris, tire fragments etc.
Requirements	:	innovation, adaptation, updating, added operational procedures and changing technology all make special new regulations and criteria a high priority

This indicates that all of the elements are involved in the appearance of mechanical damage. Our incomplete database of threats and the updating of requirements need a process for continuing case-to-case updating and incorporation into the certification requirements.

In addition to the above sampling of assignable damage there are accidental, unknown damage threat processes such as environmental extremes, unpredicted hailstorms, etc. that lie outside the design envelope.

The threats that can be characterized should have appropriate processes in place for the different elements of safety. We now will investigate the quality requirements for these processes.

10.1.2.1 Design processes and mechanical damage

Because of the steady change in materials, processes and structural concepts, empirical insights are slow to be validated. Secondary loads, local changes in geometry and gauges, complicated details and local, internal load concentration have manifested themselves as different types of defect and mechanical damage, often in locations that are not accessible to inspection and which are difficult to lay-up and process.

It would appear that the design process quality should be controlled by something of the nature of Six-Sigma supported by a historical database and including educational aims, including a degree of test validation.

If we choose a true Six-Sigma approach that focuses on 'design-caused damage' and defects outside the range of the 6σ region, then, if the quality process is properly applied, the probability of design-rooted defects is

$$P(\overline{X}_i) = P(\overline{X}_D T_i \overline{X}_t \overline{D}_{Ut}) = P(\overline{D}_{Ut} \mid \overline{X}_t T_i \overline{X}_D) \cdot P(\overline{X}_t \mid \overline{X}_D T_i) \cdot P(T_i \mid \overline{X}_D) \cdot P(\overline{X}_D)$$

$$(10.3)$$

The participating events in Equation (10.3) are

$\overline{X}_D$: defect caused by design
T_i : type of defect
$\overline{X}_t$: defect results in mechanical damage at time t
$\overline{D}_{Ut}$: damage size is larger then ultimate requirement at t.

A numerical illustration of Equation (10.3) is shown in the next example.

Example 10.4 The purpose of this example is to illustrate the orders of magnitude that practical, structural safety requirements favor:

$$P(\overline{X}_i) = 10^{-1} \cdot 10^{-2} \cdot 0.3 \cdot 10^{-8} = 0.3 \cdot 10^{-11}$$

This result seems to support the use of Six-Sigma as realistic for design defects.

10.1.2.2 Manufacturing processes and mechanical damage

Defects originating in manufacturing (processing, installation, assembly, etc.) and developing into mechanical damage are often hard to detect because of the location and placement involved.

These kinds of defect are also often regional, i.e. affecting a considerable area (many load paths), and therefore constitute a serious threat to fail-safe integrity. If the result also includes reduced structural properties such as strength, stiffness, damage resistance, damage growth rates and 'residual strength', the situation is unsafe and described by the equation given below (Equation (10.4)). The combined event describes a situation where a serious process failure renders the structural strength and damage tolerance properties below requirements, is not detected in process control or in quality control and produces mechanical damage and loss of limit integrity and failure. This combined event is described in Equation (10.4), and Example 10.5 contains an analysis that illustrates the consequences of using realistic values supporting the safety objective.

$$P(\overline{A}_{MD}) = P(\overline{X}_M T_i V_{i3} \overline{P}_C \overline{Q}_C \overline{U}_{Lt} \overline{X}_t \overline{D}_U \overline{A}_T) = P(\overline{A}_T \mid \overline{U}_{Lt} \overline{X}_t \overline{D}_U \overline{P}_C \overline{Q}_C \overline{X}_M T_i V_{i3})$$
$$\cdot\, P(\overline{Q}_C \mid \overline{P}_C \overline{X}_M T_i V_{i3}) \cdot P(\overline{P}_C \mid \overline{X}_M T_i V_{i3}) \cdot P(V_{i3} \mid T_i \overline{X}_M) \cdot P(T_i \mid \overline{X}_M)$$
$$\cdot\, P(\overline{X}_M) \cdot P(\overline{U}_{Lt} \mid \overline{X}_t \overline{D}_U \overline{X}_M T_i V_{i3}) \cdot P(\overline{D}_U \mid \overline{X}_t \overline{X}_M T_i V_{i3}) \cdot P(\overline{X}_t \mid \overline{X}_M T_i V_{i3})$$

$$(10.4)$$

The following events participate:

$\overline{X}_M$: processing defect in manufacturing
T_i : type of defect reduced strength and damage tolerance
V_{i3} : extent of damage
$\overline{P}_C$: failed process control
$\overline{Q}_C$: failure of quality control
$\overline{U}_{Lt}$: loss of limit integrity at time t
$\overline{X}_t$: mechanical damage present
$\overline{D}_U$: damage size larger than ultimate requirement
$\overline{A}_T$: failure at time T
$\overline{A}_{MD}$: failure due to manufacturing defect.

Example 10.5 The purpose of this example is to illustrate the consequences of manufacturing process failures using orders of magnitude that are consistent with composite safety requirements. Equation (10.4) is used:

$$P(\overline{A}_{MD}) = p_f^n \cdot 10^{-2} \cdot 10^{-2} \cdot 10^{-1} \cdot 0.5 \cdot 10^{-8} \cdot 0.9 \cdot 10^{-2} \cdot 10^{-1} \leq 10^{-16}$$

This result shows that this combined event has a very small probability of occurring. The quality of the manufacturing processes could be a great deal less than that which Six-Sigma yields, and this would have to be shown for many critical events. The problem wherein different materials react differently to different process defects makes it necessary to recognize that the combination of corrupted ultimate strength properties and damage tolerance characteristics is neither rare nor common and the specifics for each material system must be determined during material characterization. A reliable material-selection program must include a comprehensive test program in order to meet safety requirements.

The curing of parts and assemblies or co-bonding of major sub-assemblies requires scrutiny of process failures that introduce defects in a substantial region of a PSE and therefore cause loss of fail-safe integrity, especially when ultimate strength is reduced. For PSEs that are damage-tolerance critical, there is a built-in ultimate margin of safety that could preserve ultimate integrity for defects up to that level. However, for zero ultimate margin of safety structure, the guiding equation is

$$P(\bar{U}_{FS}) = P(\bar{X}_M T_I V_{13} \bar{U}_{U0} \bar{P}_C \bar{Q}_C M_S \bar{U}_{FS}) = P(\bar{U}_{FS} \mid M_S \bar{Q}_C \bar{P}_C \bar{U}_{U0} \bar{X}_M T_I V_{13})$$
$$\cdot P(M_S \mid \bar{Q}_C \bar{P}_C \bar{U}_{U0} \bar{X}_M T_I V_{13}) \cdot P(\bar{Q}_C \mid \bar{P}_C \bar{U}_{U0} \bar{X}_M T_I V_{13}) \cdot P(\bar{P}_C \mid \bar{U}_{U0} \bar{X}_M T_I V_{13})$$
$$\cdot P(\bar{U}_{U0} \mid \bar{X}_M T_I V_{13}) \cdot P(V_{13} \mid T_I \bar{X}_M) \cdot P(T_I \mid \bar{X}_M) \cdot P(\bar{X}_M)$$

$$(10.5)$$

The following events are involved:

$\bar{X}_M$: manufacturing defect
T_I : type of defect
V_{13} : extent of defect
$\bar{U}_{U0}$: loss of ultimate integrity
$\bar{P}_C$: not detected in process control
$\bar{Q}_C$: not detected in quality control
M_s : positive margin of safety
$\bar{U}_{FS}$: loss of fail-safe integrity.

Closer scrutiny of Equation (10.5) reveals that the first factor on the right-hand side is equal to zero if the margin of safety, M_S, 'covers' the strength reduction

$$\text{Margin in design: } MS = \frac{F_{all}}{f_{appl}} - 1 \Rightarrow F_{all} = (1 + MS)f_{appl}$$

$$\text{With defect and MS} = 0 :\Rightarrow kF_{all} = f_{appl} \Rightarrow F_{all} = \frac{1}{k} f_{appl}$$

$$\Rightarrow k = \frac{1}{1 + MS} \Rightarrow MS = \frac{1}{k} - 1$$

so the probability of lost fail-safe integrity is zero if the proper margin of safety is used.

10.1.3 Criticality and criteria

The problem with widespread processing defects that cause reduction of structural properties is that, as we have seen, loss of fail-safe integrity can result. That makes a very realistic set of design load case definitions necessary and important, allowing us to use the criticality of damage tolerance as a buffer for tolerating some reductions in ultimate strength without loss of fail-safe integrity.

The evolution of ultimate design data has followed along the following lines:

$$P(\overline{U}_U) = P(\overline{B}_U S_D S_E S_S) = P(\overline{B}_U \mid S_D S_E S_S) \cdot P(S_D) \cdot P(S_E) \cdot P(S_S) \quad (10.6)$$

Here we are using the following definitions:

$\overline{B}_U$: allowable is less than requirements
S_D : state of damage
S_E : state of environment
S_S : local internal loads disturbance.

But the process has not been consistent and we have a set of design load cases, e.g. in compression:

1. 'Compression after impact' with:
   S_D : ultimate region damage size
   S_E : room temperature, dry
   S_S : no disturbance except from damage.
2. Open hole hot wet compression with:
   S_D : no damage
   S_E : max. temperature and equilibrium moisture content
   S_S : open hole internal load concentration.
3. Buckling:
   S_D : No damage
   S_E : Room temperature
   S_S : Uniform distribution.
   Here one of these design conditions, is often reported as critical for ultimate loads. One might ask why a number of other variations are not required. Certainly in pursuing the ultimate margin of safety buffers, it is important to at least investigate the following design conditions:
4. Cold Compression after impact:
   S_D : ultimate region damage size
   S_E : $-65°F$ temperature
   S_S : uniform distribution.
5. Cold open-hole compression:
   S_D : manufacturing or repair defect
   S_E : $-65°F$ temperature
   S_S : fastener load defect due to manufacturing.

6. Cold, damaged Buckling:
    S_D : ultimate region damage size
    S_E : $-65°F$ temperature
    S_S : With or without fastener manufacturing defect in installation.
7. Actual temperature in the vicinity of an engine and air conditioning.

These suggested load cases would be part of the criticality determination and buffer determination. The comparison is between critical ultimate condition and the limit load condition with 'large' damage and a low temperature of $-65°F$.

Criticality would be established as a comparison between loss of ultimate integrity and loss of limit integrity measured in thickness.

Example 10.6 This example gives a comparison between ultimate and limit integrity. We start with a comparison between ultimate and limit integrity for an ultimate condition.

$$P(\bar{U}_U) = P(\bar{B}_U \mid S_D S_E S_S) \cdot P(S_D) \cdot P(S_E) \cdot P(S_S) \quad and$$
$$P(\bar{U}_L) = P(\bar{B}_L \mid S_D S_E S_S) \cdot P(S_D) \cdot P(S_E) \cdot P(S_S)$$

We assume normal distributions, $C_V = 0.1$ and B-values.
The ultimate B-value is

$$\Phi\left(\frac{B - \mu}{\sigma}\right) = 0.1 \Rightarrow t = -1.3 \Rightarrow \frac{B}{\mu} = 1 - 0.13 = 0.87$$

which implies a limit B-value for the same states:

$$\Phi\left(\frac{\frac{0.87}{1.5} - 1}{0.1}\right) = \Phi(-4.2) = 10^{-5}$$

And with prescribed limit states:

$$P(\bar{U}_{LD}) = P(\bar{B}_{LD} \mid S_{DL} S_E S_S) \cdot P(S_{DL}) \cdot P(S_E) \cdot P(S_S)$$

If the ratio between the probability of loss of ultimate integrity and the probability of loss of limit integrity is less than 1, a safety buffer exists that can be used to maintain fail-safe integrity.

$$R_{U/_L} = \frac{0.1 \cdot P(\bar{X}D_U)}{0.1 \cdot P(\bar{X}D_5)} = \frac{P(D_U \mid \bar{X})}{P(D_5 \mid \bar{X})} = \frac{0.9}{10^{-3}} >> 1$$

It is not surprising that preserving fail-safe integrity (preservation of ultimate integrity) is a very important design criterion because many defects encountered require a margin of survival in relation to serious, widespread defects.

10.2 CONCLUSIONS

Fail-safe integrity is a very important part of structural design and a fundamental requirement of safety management owing to the risk of widespread defects arising as a result of manufacturing processing defects.

A crucial review of what constitutes a realistic, practical set of design load cases with a realistic environmental requirement, e.g. structural panel B-values and design temperatures of $-65F$, is needed.

Safety requirements need to be based on residual strength B-values.

Chapter 11
Design Mission, Criticality and Integrity

Undetected loss of limit integrity and undetected loss of ultimate integrity (leading to loss of fail-safe integrity), can lead to very practical adaptations to design criteria or safety-based design constraints. Design loads, design states, damage probabilities and the possibility of 'walk-around' inspections vary from principal structural element (PSE) to PSE and can lead to a challenging formulation of damage tolerance and fail-safe integrity requirements. One of the more effective formulations of design criteria is PSE-based, which makes design loads formulation most effective if 'design mission-based', which also facilitates structural safety management and the control of the process defects introduced by the elements of safety. The previous ten chapters have set the stage for an informed review of orders of magnitude of the design safety requirements needed. The rest of the book investigates integrity-based design constraints and starts with damage tolerance integrity.

11.1 DAMAGE TOLERANCE INTEGRITY AND CRITICALITY

Damage tolerance critical structural designs produce a built-in ultimate margin of safety that can compensate for structural property reductions of hard-to-detect defects in terms of 'legal' requirements for Six-Sigma quality processes.

The cornerstone of this design requirement is the probability of undetected loss of integrity for design missions for one detail design point (DDP), 'i':

$$P(\bar{U}_{Li}) = \sum_{j=1}^{n_i} P(\bar{B}_{Lij} S_{Dj} S_{Ej} S_{Sj}) = \sum_{j=1}^{n_i} P(\bar{B}_{Lij} \mid S_{Dj} S_{Ej} S_{Sj}) \cdot P(S_{Dj}) \cdot P(S_{Ej}) \cdot P(S_{Sj})$$

$$(11.1)$$

The next example investigates a set of internal loads and temperatures and uses the 'probability of an unsafe state' (see Equation (11.2)) as the design constraints to consider:

$$P(\bar{S}_T) = P(\bar{U}_T \bar{H}_T \bar{A}_{\leq n})$$

$$(11.2)$$

This is a generalization of an unsafe state which includes the event, $\bar{A}_{\leq n}$, failure within n flight. Equation (11.2) in expanded form becomes:

$$P(\bar{S}_T) = P(\bar{A}_{\leq n} \mid \bar{U}_T \bar{H}_T) \cdot P(\bar{H}_T \mid \bar{U}_T) \cdot P(\bar{U}_T)$$

$$(11.3)$$

The first factor on the right hand-side of Equation (11.3) represents the probability of failure in less than n flights, given loss of damage tolerance integrity and undetected damage at time T and can be written as

$$P(\bar{A}_{\leq n} \mid \bar{U}_T \bar{H}_T) = p_f + p_s p_f + p_s^2 p_f + \cdots + p_s^{(n-1)} p_f = \frac{1 - p_s^n}{1 - p_s} p_f = 1 - p_s^n$$

Here p_f is probability of failure given loss of damage tolerance integrity, and p_s is the probability of survival given loss of damage tolerance integrity.

The order of magnitude of Equation (11.3) for $n = 3000$ and residual strength B-values is:

$$P(\bar{S}_T) \approx 0.95 \cdot 10^{-3} \cdot 10^{-6} \approx 10^{-9}$$

This order of magnitude illustrates the role of scheduled inspections in safety.

The next example describes the use of 'design mission' as a way of introducing safety constraints into the design process.

Example 11.1   The purpose of this example is to demonstrate practical use of internal design loads and the selection of different environments and damage scenarios. The situations are different at different locations. At locations where walk-around inspections are possible, the probability of detection is relatively modest. We start with a comparison between probability of detection and probability of survival of n flights.

The probability of survival of n is (based on a typical design):

$$p_{sn} = 0.9997^n$$

and the probability of detection in n flights, assuming a probability of detection for a random flight of 0.5, is:

$$p_{dn} = 1 - 0.4998^n$$

The following table gives the comparison for a realistic number of flights:

n flights	Probability of surviving	Probability of detecting damage
10	0.997	0.999
20	0.994	0.999999
30	0.991	~1
40	0.988	1

This comparison shows that even with a probability of detection not larger than 0.5, walk-around inspections are very powerful contributors to safety.

The next example describes the use of Equation (11.3) for designing to a prescribed minimum probability of an unsafe state.

Example 11.2   This example deals with $n = 3000$ and a set of internal design loads and a set of different environments based on the fact that many PSEs are exposed to outside temperatures or to elevated temperature from one source or an other. Thus a DDP can have the following design mission definition:

#	Internal response	Damage	Environment	Disturbance
1	L_1	D_5	$-65F$	OH
2	L_1	–	T_{MAX}	OH
3	L_1	–	$-65F$	Manuf. defect
4	$0.9L_1$	–	$-65F$	OH
5	$0.8L_1$	–	$-65F$	OH

This set of scenarios will now be investigated using the equation

$$P(\bar{A}_n \mid \bar{U}_T \bar{H}_T) \cdot P(\bar{H}_T \mid U_T) \cdot P(\bar{U}_T)$$

where the probability of loss of integrity is

$$P(\bar{U}_T) = P(\bar{B}_{LT} \mid S_D S_E S_S) \cdot P(S_D) \cdot P(S_S)$$

The cases are:

case 1: $0.95 \cdot 10^{-3} \cdot 10^{-1} \cdot 10^{-5} \cdot 0.6 \cdot 0.1 = 0.6 \cdot 10^{-10} = 6 \cdot 10^{-11}$
case 2: $0.95 \cdot 10^{-3} \cdot 10^{-1} \cdot 10^{-5} \cdot 10^{-3} \cdot 10^{-1} = 10^{-13} = 0.01 \cdot 10^{-11}$
case 3: $0.95 \cdot 10^{-3} \cdot 10^{-1} \cdot 10^{-5} \cdot 0.6 \cdot 0.8 \cdot 10^{-9} = 0.4 \cdot 10^{-18} \sim 0$
case 4: $0.95 \cdot 10^{-3} \cdot 1.4 \cdot 10^{-2} \cdot 10^{-5} \cdot 0.6 \cdot 10^{-1} = 0.8 \cdot 10^{-11}$
case 5: $0.95 \cdot 10^{-3} \cdot 1.4 \cdot 10^{-3} \cdot 10^{-5} \cdot 0.6 \cdot 10^{-1} = 0.8 \cdot 10^{-12} = 0.1 \cdot 10^{-11}$
Total: $0.7 \cdot 10^{-10}$

A total for the PSE (25 DDPs): $25 \cdot 0.7 \cdot 10^{-10} = \mathbf{1.8 \cdot 10^{-9}}$

If we now assume that the requirement for this case PSE is $2 \cdot 10^{-8}$, which per DDP would be $0.8 \cdot 10^{-9}$

$$\text{requirement } 0.8 \cdot 10^{-9} \geqslant \text{ actual value } 0.7 \cdot 10^{-10}$$

So thicknesses could be reduced so that a factor of ~9 is obtained:

Case 1 becomes $0.7 \cdot 10^{-11}$ (yields 1/9 of 0.10), and the allowable value for

$$t = -2.32 = \frac{d - \mu}{\sigma} \Rightarrow \frac{d}{\mu} = 1 - 0.2 = 0.77 \Rightarrow \text{ guage reduction } \frac{0.77}{0.87} = \sim 0.89$$

Case 2 and 3 is sufficiently small
Case 4 becomes $0.2 \cdot 10^{-11}$ (yields 1/9 of 0.014), and the allowable value for

$$t = -3.68 = \frac{d - \mu}{\sigma} \Rightarrow \frac{d}{\mu} = 0.63 \Rightarrow \text{ guage reduction } \frac{0.63}{0.70} = 0.90$$

Case 5 becomes 10^{-11} (yields 1/9 of 0.0014), and the allowable value for

$$t = -4.20 = \frac{d - \mu}{\sigma} \Rightarrow \frac{d}{\mu} = 0.58 \Rightarrow \text{ guage reduction } \frac{0.58}{0.64} = 0.91$$

This approach is easily automated and the required inputs are design values, design loads and 'initial' guages. The results include required guages and structural safety values, given that the processes for the other elements of safety are defects-free.

11.2 CRITICALITY AND PROCESS DEFECTS

Case 3, in the previous section, is an order of magnitude demonstration of an unadulterated Six-Sigma application and the result provides much greater safety than needed, by several orders of magnitude. Considering the difficulty in establishing and maintaining 'tight requirements', it seems a reasonable starting point for some processes to accept a starting point that agrees with the drift we presently tolerate, i.e. $\Phi\,(-4.5)$.

The previous numerical investigations have given a good review of what difficult situations throw up from a safety standpoint. Manufacturing in particular has to deal with defects in processing that are especially challenging. Defects that reduce structural properties can be widespread and hard to detect.

We have just seen that reduction in residual strength is far from difficult to handle. The reduction of ultimate strength due to defects, on the other hand, is very crucial, because of its close relationship to fail-safe design. The use of B-allowable values requires fail-safe design, or whatever modern terminology chooses to call it these days. So, in order to benefit from B-values in engineering, you must have a multi-load-path structure that can carry internal, ultimate loads or more when load-path failure causes remaining structure to become overloaded (larger than ultimate). Widespread reduced ultimate strength without an initial positive margin of safety buffer is an unacceptable state that must be identified by process control or quality control.

If this were to happen in residual strength critical structure, there would be a built-in ultimate margin that would preserve fail-safe integrity up to the limit of that margin. The next example illustrates the range of a reasonable margin of safety.

Example 11.3　We assume that the residual strength is approximately proportional to the square root of the damage size. The damage regions are defined in Figure 11.1.

We now assume that ultimate thickness would be based on a damage size of L; and the thickness would be:

$$t_U = \frac{N_U}{RS_U} \Rightarrow \frac{N_U}{t_U\,RS_U} = 1$$

The thickness based on limit requirements would depend on a damage size of 3.0 L (shown in Figure 11.1), and a square-root relation between different damage sizes, yielding

$$t_L = \frac{N_U\sqrt{3.5}}{1.5 RS_U} = 1.25\,\frac{N_U}{RS_U}$$

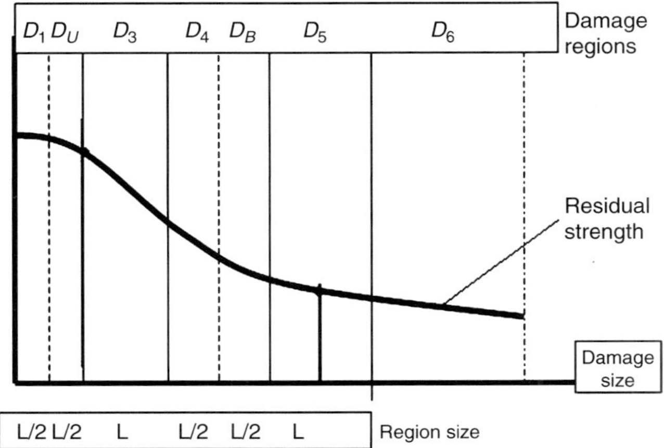

Figure 11.1 Example of damage size regions.

Thus the ultimate margin of safety is

$$\frac{N_U}{t_L F_U} = \frac{N_U}{1.25 t_U F_U} \Rightarrow \text{reduced ultimate allowable value '}k\text{'}$$

$$\frac{N_U}{1.25 t_U k F_U} \Rightarrow k = \frac{1}{1.25} = \frac{1}{1+MS} = 0.80$$

where the process quality could be shown, as in Figure 11.2, to be 4.5σ, and the fact that a good damage tolerance strategy can be used as a buffer for defects that reduce ultimate strength up to 20%; e.g. if circumstances support and preserve fail-safe integrity, especially if practical constraints do not necessitate more stringent measures.

The nature of these types of process defect is widespread and hard to detect. A safety solution for this scenario would deal with both, and fail-safety would compensate for concurrent reduced residual strength.

A design strategy that does preserve ultimate integrity even after process failures that reduce ultimate strength and its consequences is a very important safety feature if quality control processes and other processes involved are to have a very high probability of success.

The recovery and critical joint events are

$$P(S_M) = P(\bar{X}_M \bar{P}_C Q_C) \cup P(\bar{X}_M P_C) \cup P(X_M)$$

and

$$P(\bar{S}_M) = P(\bar{X}_M \bar{P}_C \bar{Q}_C) \tag{11.4}$$

Equation (11.4) can be interpreted as describing the relation between safety and 'generalized manufacturing success'. The following events participate:

$\bar{X}_M$: defective manufacturing process
P_C : successful process control
Q_C : successful quality control.

The first version of the equation describes success in manufacturing with passive process control.

The first term represents rescue by quality control but with rejection;
The second term represents detection by the process control (passive) and rejection;
The third term represents a defect–free process.

The second version of Equation (11.4) represents a generalized manufacturing failure.

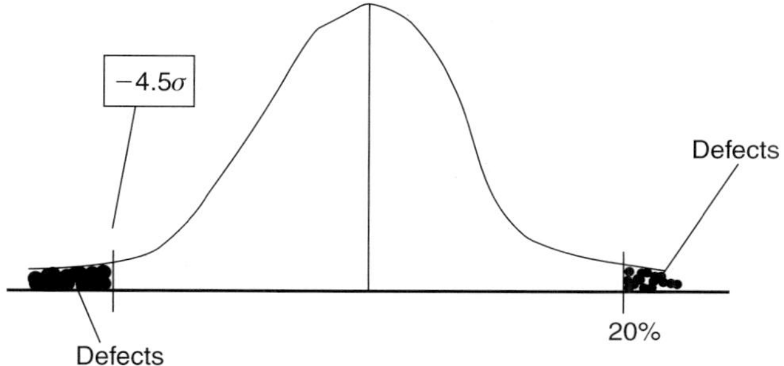

Figure 11.2 Quality 4.5 Sigma strength reductions unacceptable at 20%.

Applying this definition often results in expensive failures. If however, an active process control could be introduced there would be significant cost reduction. It is also easy to understand the allure of thermoplastics, which would produce success by reprocessing.

The first version of Equation (11.4) can provide an avenue for describing orders of magnitude.

We will compare versions of the quality description shown in Figure 11.2. An expanded version of Equation 11.4 will be used.

$$
\begin{aligned}
P(S_M) &= P(\bar{X}_M \bar{P}_C Q_C) + P(\bar{X}_M P_C) + P(X_M) \\
&= P(Q_C \mid \bar{P}_C \bar{X}_M) \cdot P(\bar{P}_C \mid \bar{X}_M) \cdot P(\bar{X}_M) \\
&\quad + P(P_C \mid \bar{X}_M) \cdot P(\bar{X}_M) + P(X_M) \\
&= 1 - P(\bar{Q}_C \mid \bar{P}_C \bar{X}_M) \cdot P(\bar{P}_C \mid \bar{X}_M) \cdot P(\bar{X}_M)
\end{aligned}
\tag{11.5}
$$

and

$$
P(\bar{S}_M) = P(\bar{Q}_C \mid \bar{P}_C \bar{X}_M) \cdot P(\bar{P}_C \mid \bar{X}_M) \cdot P(\bar{X}_M)
$$

The second version of Equation (11.5) will then be used to evaluate requirements for a widespread failure of a PSE; the requirement becomes

$$
P(\bar{S}_M) \leqslant 10^{-9}
$$

This specific requirement (e.g. given airplane safety requirement, structure's share and 50 PSEs) would be satisfied by the case of equal contributions from all elements, resulting in equal factors in Equation (11.5).

The requirement would be satisfied by equal factors in Equation (11.5), which would require:

$$
P(\bar{S}_M) = (10^{-3})^3 = 10^{-9}
$$

If we instead were to apply the quality shown in Figure 11.2 to the basic manufacturing process quality, we get:

$$P(\overline{S}_M) = P(\overline{Q}_C \mid \overline{P}_C \overline{X}_M) \cdot P(\overline{P}_C \mid \overline{X}_M) \cdot 3.4 \cdot 10^{-6} \leqslant 10^{-9}$$
$$\Rightarrow P(\overline{Q}_C \mid \overline{P}_C \overline{X}_M) \cdot P(\overline{P}_C \overline{X}_M) \leqslant 3 \cdot 10^{-4}$$

The result is:

$$P(\overline{Q}_C \mid \overline{P}_C \overline{X}_M) = P(\overline{P}_C \mid \overline{X}_M) \leqslant 1.7 \cdot 10^{-2}$$

The above result represents approximately a quality of 2σ. As the situation represents a hard-to-detect situation it does not seem to be unreasonable. It is also worth noting that this situation also changes from case to case (e.g. change in composite material, process and structural concept).

11.3 CRITICALITY AND REPAIR PROCESSES

This review of manufacturing process defects has identified three characteristics that challenge the safety of composite structure. Defects are often widespread and hard to detect, especially after service has been entered, which, when the results involve reduction in structural strength, compromises fail-safe integrity and challenges damage tolerance integrity and inspection effectiveness.

This context requires a structural safety management approach that deals with both widespread and local defects through the use of high-quality processes (including in structural design), and a very balanced 'generalized manufacturing process', which focuses on minimizing MRB rejections, production repairs and life-cycle costs.

Process qualities identified above and structural integrity preservations are but a beginning. A dedicated application of safety management is imperative for all reliable combinations of innovation and safety.

Repair in service has similarities to manufacturing in the context of process defects, but takes place under much more difficult circumstances. The controlling equation is for the probability of an unsafe state:

$$P(\overline{S}_T) = P(\overline{X}_t D_{5t} \overline{B}_{Lt} H_T R_T \overline{X}_{TI} T_1 \overline{U}_{UT})$$
$$= P(\overline{U}_{UT} \mid \overline{X}_{TI} T_1) \cdot P(T_1 \mid \overline{X}_{TI}) \cdot P(\overline{X}_{TI})$$
$$\cdot P(R_T \mid H_T \overline{B}_{Lt} D_{5t} \overline{X}_t) \cdot P(H_T \mid \overline{B}_{Lt} \overline{X}_t D_{5t})$$
$$\cdot P(\overline{B}_{Lt} \mid \overline{X}_t D_{5t}) \cdot P(D_{5t} \mid \overline{X}_t) \cdot P(\overline{X}_t) \tag{11.6}$$

The participating events are:

$$\overline{X}_t \quad : \quad \text{damage present (or damage larger than sizes contained in } D_3\text{)}$$
$$D_{5t} \quad : \quad \text{damage size at ``}t\text{'' belong in limit requirement range}$$
$$\overline{B}_{Lt} \quad : \quad \text{residual strength less than requirement limit value}$$
$$H_T \quad : \quad \text{damage detected at } T$$
$$R_T \quad : \quad \text{repair completed at } T$$
$$\overline{X}_{TI} \quad : \quad \text{repair process failure at } T$$
$$T_1 \quad : \quad \text{bonding process failure}$$
$$\overline{U}_{UT} \quad : \quad \text{loss of ultimate integrity.}$$

Example 11.4 This example investigates order of magnitudes of the probability of an unsafe state by evaluating Equation (11.6)

$$P(\overline{S}_T) = 0.10 \cdot 0.3 \cdot 0.7 \cdot P(\overline{X}_{TI}) \cdot 0.6 \cdot 0.999 \cdot 0.10 \cdot 10^{-3} \cdot 10^{-2} = \cdot 10^{-8} \cdot P(\overline{X}_{TI})$$

which for a bonding failure and 50 DDPs yields a requirement of:

$$P(\overline{S}_T) = 10^{-8} \cdot P(\overline{X}_{TI}) \leqslant 4 \cdot 10^{-10} \Rightarrow P(\overline{X}_{TI}) \leqslant 4 \cdot 10^{-2}$$

and with an order of magnitude, more stringent safety requirement:

$$P(\overline{X}_{TI}) \leqslant 4 \cdot 10^{-3}$$

A modest quality requirement for repair bonding processes would be

$$1.75\sigma \quad \text{or} \quad 2.65\sigma$$

This could also be adapted to a two-phase design with a bolted repair substructure, with total limit capability and bonded closure combining to satisfy ultimate requirements.

11.4 CRITICALITY AND DAMAGE TOLERANCE RELATED PROCESSES

The damage tolerance sub-element of safety process is a generalization of three damage safety processes:

1. damage resistance
2. damage growth resistance
3. residual strength with load-path failed.

Reduction of structural properties can involve one or any of these topics.

Damage resistance is a very important property for composite structure. It is important from the standpoint of surviving load-path-failure under load. It is also imperative in prolonging survival until the failed load-path is detected.

The limiting property is damage resistance, as the structure must redistribute internal loads to survive failure of one load-path. However, a reduction of residual strength by 20% can result in a low number of flights for only minor changes in the design mission. So, material-related systems for which the three damage safety processes degrade together due to processing defects, should be avoided otherwise residual strength reductions greater than 10% in the structure must not enter service.

11.5 MECHANICAL DAMAGE THREATS

Mechanical damage such as de-laminations, de-bonds, fiber disruptions, crushing, etc. has many causes and often the origins lie in defects associated with the elements of safety. Many defects originating with the elements are difficult to detect and conversion to mechanical damage facilitates detection.

Sub-processes exist that are associated with the elements of safety that in service produce mechanical damage. The elements involved are:

- manufacturing processes such as fastener installation defects and bonding defects;
- maintenance defects, scheduled maintenance failure, accidents in accessing target sites;
- design defects;
- operational defects including accidental, unreported damage;
- defects in the requirements definition;
- inspection mistakes.

A unified process quality for all of these could produce a widespread reduction in the uncertainty of the presence of damage threats.

As the probability of damage being present is a central part of structural monitoring in service, it would be constructive to focus on damage in the size region

$$\bar{X}_t = \bar{X}_t(\bar{D}_U) = \bar{X}_t(D_3 \cup D_4 \cup D_B \cup D_5 \cup D_6)$$

the focus being on detectable damage.

Thus, a definition of the quality of the subject processes could be initiated as a 4.5σ and the probability of defects that cause mechanical damage in service could be expressed as

$$P[\bar{X}_i(\bar{D}_U)] = P(\bar{X}_{Ei}T_1V_{13}) = P(V_{13} \mid T_1\bar{X}_{Ei}) \cdot P(T_1 \mid \bar{X}_{Ei}) \cdot P(\bar{X}_{EI}) \qquad (11.7)$$

Example 11.5 This example illustrates the order of magnitude of the probability of damage being present, and Equation (11.7) produces the 4.5σ starting point as

For one DDP: $0.5 \cdot 0.2 \cdot 0.33 \cdot 10^{-5} = 0.33 \cdot 10^{-6}$
For the PSE (50 DDPs) $= 0.165 \cdot 10^{-4}$
For 6 sub-processes $= 10^{-4}$

If we instead attempt a 3.4σ, we get for one DDP: $0.5 \cdot 0.2 \cdot 0.32 \cdot 10^{-3}$, and for 6 sub-processes

$$P[\bar{X}_i(\bar{D}_U)] = 10^{-2}$$

11.6 CONCLUSIONS

The identification of a PSE $=$ specific design mission is valuable for both providing a 'buffer' for widespread defects and yielding damage tolerance critical structure.

It is also essential for producing realistic limit damage size requirements.

All elements of safety defects contribute to mechanical damage.

A successful monitoring process requires practical data formats to contribute to reductions in risk and uncertainty.

Chapter 12
Risk and Uncertainty

Chapter 9 sets the stage for monitoring defects and performance of composite structure in an environment of innovation. The overall objective of structural monitoring is to contain risk and reduce uncertainty. The principal attack on risk is launched based on detected (in both scheduled inspections and walk-around pre-flight inspection when possible) defects, flaws and damage. Uncertainty is reduced by using collected data to update a priori probability distributions. The basic data are originally PSE-specific and can, when similar situations for several PSEs have been established, be extended to a groups of PSEs. An accumulative database in time can also improve the quality of safety.

The target data include: probability of damage being present, the probability of their being in a specific size region and the probability of detection. Figure 12.1 describes a generic arrangement of size regions that can become specific when design criteria are created. This representation can be the basis for a distribution function of residual strength, given detected damage, $F(RS|D_sH)$. Figure 12.2 describes the type of results that can be obtained (see also Feller 1957).

The main tool for risk reduction is adjusting inspection periods and approaches so that the probability of an unsafe state can be kept lower, based on inspection data gathered.

Updated probability of damage present per PSE will be collected in order to understand differences between PSEs. The uncertainty will be reduced through the updating of probability distributions for damage sizes.

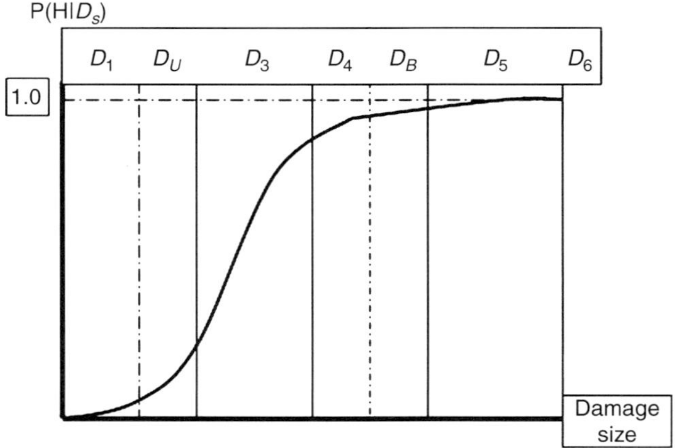

Figure 12.1 Detection versus damage size.

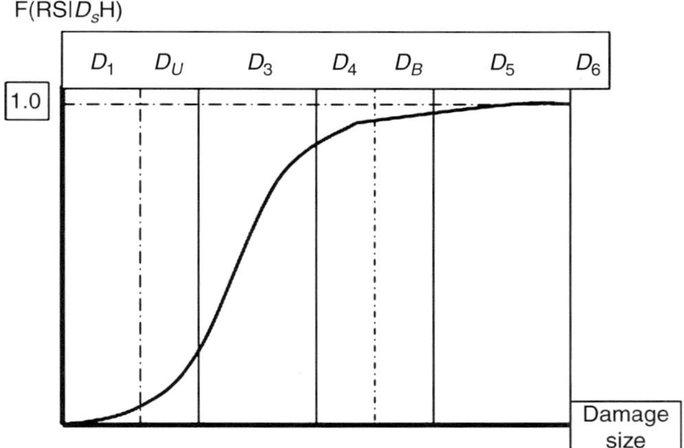

Figure 12.2 Distribution of residual strength; damage detected.

The following size regions are included in the two figures:

D_1 : undetectable damage-size
D_U : ultimate damage region
D_3 : transition region
D_4 : damage resistance design region
D_B : damage growth region
D_5 : limit load design region
D_6 : excessive damage region.

Figure 12.2 describes the distribution of residual strength for detected damage and will be used for uncertainty reduction and the foundation for building the service database and the baseline allowable values.

In identifying the markers to use to activate the risk management we assume that a complete set of values for the size regions has been developed for Figures 12.1 and 12.2 and fail-safe criteria are in place. Discovery of damage in region D_5 and D_6 or failure of a load-path should activate a close-up scrutiny of the location. Special investigations of the detected sites will be called for other airplanes. At the same time there is an ongoing validation that widespread manufacturing defects are not present.

For situations when walk-around inspections are not possible, we will focus on the probability of generalized unsafe states. The equation describing this probability is

$$P(\bar{S}_t) = P(\bar{U}_T \bar{H}_T \bar{A}_n) = P(\bar{A}_n \mid \bar{H}_T \bar{U}_T) \cdot P(\bar{H}_T \mid \bar{U}_T) \cdot P(\bar{U}_T) \qquad (12.1)$$

Example 12.1   We assume original values and 25 DDPs, which yields

$$n \qquad P(\bar{S}_t) = P(\vec{A}_n \mid \bar{H}_T \bar{U}_T) \dots \qquad \text{next inspection}$$

$$1 \qquad\qquad - \qquad\qquad\qquad\qquad -$$

$$1000 \qquad 0.74 \cdot 10^{-9} \qquad\qquad\qquad 1000 \text{ flights}$$

The probability that the structure will survive 1000 flights is 74%, which is for B-value residual strength. The value should be prescribed in a future composite regulation.

12.1   MARKERS FOR MONITORING ACTION

The detail requirements depend on the material, on the processes involved and on the structural concepts and the timing. For this discussion we assume the following set of requirements and consequential criteria. The probabilities of an 'unsafe flight' are:

For the airplane 10^{-5}
For the structure 10^{-6}
For each element of safety (assumed equal) $2 \cdot 10^{-7}$
For each PSE (e.g. 100) $2 \cdot 10^{-9}$
For each DDP (e.g. 100) $2 \cdot 10^{-11}$

The expectation is that the probability of an unsafe state after inspection should be less than requirements. An illustration of order of magnitude is given in the next example.

Example 12.2   This example deals with two defects in region 5, D_5, only one being detected. The probability of this situation is

$$P(\bar{S}_P) = \left[\binom{20}{1} \cdot P(\overline{UH}_T) \right] \cdot \left[\binom{19}{1} \cdot P(\overline{UH}_T) \right] \cdot \left[\binom{18}{1} \cdot P(\overline{UH}_T) \right]$$
$$= 20 \cdot 19 \cdot 18 \cdot (10^{-3} \cdot 10^{-1} \cdot 10^{-3} \cdot 10^{-2})^3 = 6840 \cdot 10^{-27}$$

Only one missed in D_5

$$P(\bar{S}_P) = 20 \cdot 10^{-3} \cdot 10^{-1} \cdot 10^{-3} \cdot 10^{-2} = 2 \cdot 10^{-10}$$

Only one missed in D_B out of 20

$$P(\bar{S}_P) = 20 \cdot 10^{-2} \cdot 0.0026 \cdot 10^{-2} \cdot 10^{-2} = 0.5 \cdot 10^{-7} \text{ which is unsafe!}$$

The last value does not satisfy the requirement; because it is not below $2 \cdot 10^{-9}$ (even if the number of defects were considerably lower), this situation should trigger a second intense inspection to make sure that there is no 'large' damage remaining undetected.

The following marker must lead to action to ensure acceptable level of safety through the next inspection period. In addition to no damage in D_5, damage in D_6 must be reported to engineering for assessment of the inspection period and approach. The inspection results must also be used for updating the probability distributions for damage and detection.

The failure of a load-path must be identified and repairs completed before the airplane can fly again, because additional partial failures can occur, within just a few flights, if this is not done.

12.2 SITUATIONS AFTER SCHEDULED INSPECTIONS

A PSE failure is critical, especially when walk-around inspections are not realistic, and the efficacy of scheduled inspections are all-important. The next example illustrates how a relatively difficult situation can prevail after a major inspection.

Example 12.3   We assume that we have a PSE with one hundred DDPs and that the design model (the DDP configuration) satisfies the internal loads distribution, the sizing requirements and the definition of damage locations. The following equation describes the situation after a major inspection when all damage is discovered and repaired, except for three large defects. The probability of an unsafe state for the first flight after inspection is, assuming all defects are in D_4, (9), in D_B, (7), and in D_5, (4), one in each region being missed. The situation before the inspection is described by the following equations:

$$P(\overline{S}_P) = 9P(\Delta \overline{S}_{D4}) + 7P(\Delta \overline{S}_{DB}) + 4P(\Delta \overline{S}_{D5})$$

The probability of one member 'missed' in: $D_5 = 0.4 \cdot 10^{-8}$
$$D_B = 1.8 \cdot 10^{-8}$$
$$D_4 = 0.5 \cdot 10^{-8}$$

So either of these three cases is unsafe because the bound is $2 \cdot 10^{-9}$.

12.3 WIDESPREAD DEFECTS

Process failure in manufacturing (such as part processing or co-bonding), can extend over a substantial part of a PSE. The region could have lost fail-safe integrity with defects hard to detect, and, once the airplane enters service, detection may not be an option. The concept of acceptable defects lends itself to being addressed when limit damage regions are determined. If the region is such that residual strength is critical and there is an ultimate strength margin of safety, it can be used as a buffer for ultimate strength reductions and could for example be of the order of magnitude of 20%, which could be included under the Six-Sigma or equivalent quality definitions.

The monitoring/analysis process could use a comparison algorithm for probability of defects being present for the same PSE for different airplanes, the accumulated number of defects and repairs for a specific PSE, and the probability of large damage sizes. The purpose is to find 'unusual defect' patterns. The comparisons could be extended to data from different airlines under the auspices of the FAA and other international agencies. Material systems that could be expected to display both ultimate strength reduction and damage tolerance degradation should be avoided or subject to very stringent process control and quality control.

The monitoring/analysis process includes data for updating, comparisons, verification and diagnosis, all for preserving or restoring the level of safety. The monitoring, updating and improving of damage accumulation requires a process that can track the origin of the damage. All elements are contributors, and the development of the initial process must have a measurable foundation.

12.4   INITIAL DAMAGE REGIONS AND PROCESS DEFINITIONS

The process could for example be treated as a 6σ process; Figure 12.3 describes the philosophy behind this.

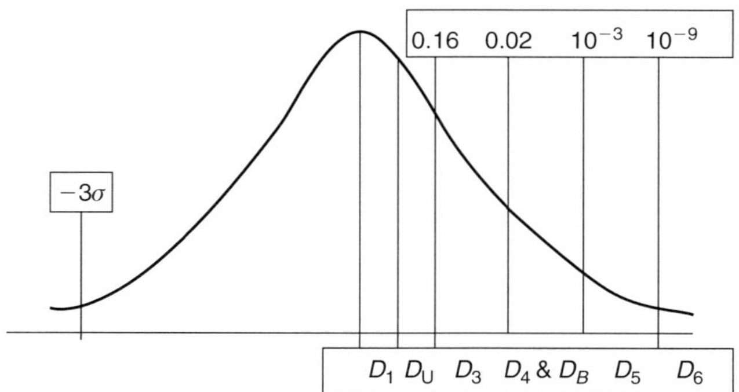

Figure 12.3 Probability distribution of damage size regions.

Figure 12.3 describes a comparison between damage size regions and the following probabilities in 6σ terms:

Region	Probability contents	Sigma interval (vertical lines)
D_1	0.40	0–0.5σ
D_U	0.36	0.5σ–σ
D_3	0.16	σ–2σ
D_4 & D_B	0.02	2σ–3σ
D_5	10^{-3}	3σ–6σ
D_6	10^{-9}	$>6\sigma$

All the elements of safety contribute to these mechanical damage sizes and it's expected that each process controlling damage accumulation and reporting has 6σ quality and is monitored with that expectation, and the personnel involved are able to compare their records in inflicting and repair damage.

12.5 CONCLUSIONS

Updating of all probability distributions, is necessary;

Comparisons between different airplanes are essential to find undetected defects;

Cumulative count of large damage, high probabilities, a high number of repairs and locally high probabilities of larger damage is very important to report.

Change in damage frequencies and size should be reported.

Chapter 13
Elements of Safety and their Characteristics

The numerical investigation of the different effects and threats for the 'elements of safety' from the standpoint of structural safety in general and structural integrity in particular is summarized in this chapter. The investigation also reveals that the structural integrities discussed below must be preserved throughout the life of the airplane structure. The following chapter contains detailed summaries of the different elements of safety.

13.1 ELEMENT OF STRUCTURAL DESIGN

The objective of this element is (1) to design the total structure in terms of PSE-specific requirements and conduct detail design in accordance with active specification and current regulations and (2) to demonstrate success through a complete test program designed to validate the design from the standpoint of ultimate integrity, limit integrity and fail-safe integrity for defects identified as 'legal' in the specifications being integrated into design missions for different PSEs.

The objective also involves including design features that facilitates preserving critical integrity. Thus damage tolerance critical design should be an objective, achieving a positive ultimate margin of safety which would allow defects up to a certain level, without violating failsafe integrity. This would also be particularly important when dealing with widespread defects caused by mishaps in basic processing, even though rare events. If this 'tactical' margin of safety (e.g. 25% would allow a 20% reduction in strength properties without loss of fail-safe integrity), were to be part of the design requirements, the unsafe state would not be reached, and, instead, a standby fail-safe capability would render widespread defects manageable, even if structural, ultimate and damage tolerance properties are reduced. The alternative to this type of design would be a process quality that yields a situation of exceedingly low probability of this type of mishap. Widespread defects that cause an unsafe state can be written as

$$
\begin{aligned}
P(\bar{S}_M) &= P(\bar{X}_M T_1 V_{13} \bar{U}_U) = P(\bar{U}_U \mid \bar{X}_U T_1 V_{13}) \cdot P(V_{13} \mid T_1 \bar{X}_M) \\
&\quad \cdot P(T_1 \mid \bar{X}_M) \cdot P(\bar{X}_M) \\
&= 0.1 \cdot 0.33 \cdot 0.1 \cdot 0.34 \cdot 10^{-5} \approx 10^{-8} \qquad \text{for } 4.5 \cdot \sigma \text{ or} \\
&= 0.1 \cdot 0.33 \cdot 0.1 \cdot 10^{-9} = 0.33 \cdot 10^{-11} \qquad \text{for } 6.0 \cdot \sigma
\end{aligned}
$$

This yields about 5.0σ for an acceptable probability of an unsafe state, but could still give in drastic results due to unexpected coincidences (e.g. discrete load accidents). It seems that a design strategy that makes damage tolerance critical as opposed to ultimate, is a vital design consideration that also promotes fail-safety.

The defects that threaten safety levels most severely are widespread. Reduction of ultimate strength properties beyond a particular value protected by the quality of the process must be 'missed' in process control and quality control with very low probability because 'significant' damage by definition is present. Among these are the reduction of strength and stiffness, as discussed above. The process quality required to produce an acceptable probability of unsafe states is ensured by making damage tolerance the critical design mode.

Reduction of damage tolerance related properties, i.e. damage resistance, damage growth rates and residual strength, is sometimes likely to happen together with reduction of ultimate properties, in which case a margin of safety buffer due to the damage tolerance defect size requirements can be the only solution and has to be checked by monitoring and accumulating globally shared databases.

For situations when only damage tolerance properties are compromised, widespread fail-safe integrity is the only saving grace. However, the probability of surviving 100 flights is just above 0.5, so detection in process control and quality control is of utmost importance in defining requirements in process control.

Defects causing built-in loads due to installation and assembly or geometry defects producing local load concentrations only become threats when combined with high flight loads. Therefore only a small number of cycles occur in an inspection period and an unsafe state is very improbable as is any threat to fail-safe integrity.

13.2 ELEMENT OF MANUFACTURING

The previous section illustrates the interaction between structural design and defects caused by mishaps in manufacturing and also the importance of including maximum effects attributable to defects within the 'legal' range of the quality definition of all manufacturing processes.

The importance of a high-quality process, process control and quality control was also emphasized in order for 'unsafe defects' to be present when the airplane enters service.

The development of manufacturing processes, such as curing and co-bonding of parts or assemblies must be jointly evolved with quality control processes and the combination of robust basic operations and 6σ process, along with quality control guardianship.

The threats to structural design from manufacturing come in many forms of different degree, starting with permanent, widespread, defective structural properties, and defects that are local and temporary and cause exceedances of local internal limit loads, or mishaps causing local undetected mechanical damage. All of these are very important for structural safety and design, but in no way do concerns over safety belong in the sphere of manufacturing alone.

13.3 ELEMENT OF MAINTENANCE

Defects in maintenance affecting structural safety should be considered in structural design including scheduled maintenance.

Inadequate scheduled maintenance causing internal load increases (e.g. jack-screw greasing): process quality should allow for some load increase within the 'legal' range.

We assume normal distribution for strength and load at attachments. If we allow for the increase to be 20%, then the survival probabilities are for example:

n	Probability of surviving n flights
1	0.986
100	0.24

Design to 20% higher load:

n	Probability of surviving n flights
1	0.99977
1000	0.79
3000	0.50

The desirable process quality for scheduled maintenance will determine inclusion in design requirements:

$$P(\overline{S}_I) = P(\overline{U}_I \overline{X}_I T_1 V_{13}) = P(\overline{U}_I \mid \overline{X}_I T_1 V_{13}) \cdot P(V_{13} \mid T_1 \overline{X}_I) \cdot P(T_1 \mid \overline{X}_I) \cdot P(\overline{X}_I)$$

The result for 6σ (20% load increase not included in the design):

$$P(\overline{S}_I) = 0.67 \cdot 10^{-1} \cdot 10^{-1} \cdot 10^{-9} = 0.67 \cdot 10^{-11} \le 4 \cdot 10^{-11}$$

Mishaps in inspection and reporting activities cause the introduction of 'large' damage into safety-related critical structure and difficult repairs result when fail-safe activates. Updating of the probability of detection and occurrence of large damage in service yields a warped picture of reality, so preserving reliable process quality is very important for safety, both in the present and the future.

Repair processes in maintenance produce the same type of defects as in manufacturing and in MR & B actions, but under more difficult circumstances. When it comes to the defects that cause reduced strength due to processing mishaps, a repair should be designed fail-safe or subject to special inspections, even though it exists for a shorter time than for undetected manufacturing defects.

Operational organizations have responsibility for ground operation, flight, walk-around inspections, incidence reporting and avoidance of severe environments (outside envelope).

The most important aspect is defects that leave permanent and clear imprints and which are missed in walk-around inspections, or observed but not reported especially in relation to serious ground damage. These are easily preventable and pre-flight inspections in particular, if doggedly pursued, can yield high detection probabilities, without falling short of 'high-tech' methods. Walk-around inspections are very effective even with modest detectibility at a random flight. A numerical illustration with probability of detection, $p_D = 0.5$, and a probability of survival $p_S = 0.9$ yields the following results:

n flights	Detection in less than n flights	Probability of failure in n flights
5	0.98	0.41
10	0.9997	0.65
20	~1	0.82
30	1	0.96
40	1	0.985

These results favor the likelihood of detection; walk-around inspections are a good measure to implement.

13.4 ELEMENT OF REQUIREMENTS FORMULATION

The last of the elements of safety mentioned, namely requirements formulation and the adaptation thereof to many new and changing needs, very much characterizes the world of composites. The importance of process failures and defects opens a variety of challenges in process development, material characterization and structural design.

The investigation of design strategy has been discussed in previous chapters. The promise of processes like 6σ was demonstrated in process characterization and development. There has also been a tendency to characterize aerospace structures as tension structures, compression structures and shear structures. However, internal load reversals are common. The 'tension' structures of wings are known to also carry up to 60% compression under negative maneuver, and some significant negative loads can occur during taxi and dynamic landing. Empennages and fuselages often experience total load reversals due to the nature of critical load cases. Reversals of internal loads such as 'stall buffet' also complicate the effects of defects as such effects are often found to be more severe in compression than in tension. It has also been found that the reductions of shear properties are often the first to develop and become most significant, involving ultimate shear strength, shear stiffness and shear strain energy rates.

Different material systems have different defects, which renders material selection and characterization a very challenging quest. Safety management requires a detail evaluation of critical effects and a consistent set of allowable values and design data, and detailed regulations are needed.

The resulting process quality, including the effects of defects, must be tightly controlled to protect safety. Ultimate integrity can be expressed as the sum integrity in tension, compression and shear:

$$P(U_U) = P(U_{UT} \cdot U_{UC} \cdot U_{US}) \approx 1 - (P(\bar{U}_{UT}) + P(\bar{U}_{UC}) + P(\bar{U}_{US}))$$
$$P(\bar{U}_U) = P(\bar{U}_{UT}) + P(\bar{U}_{UC}) + P(\bar{U}_{US}) \tag{13.1}$$

The bugaboo here is consistency of compared design values; otherwise, an objective criterion is not obtained. The format often displayed in the composites world is

$$P(\bar{U}_{UI}) = P(\bar{B}_{UI} S_D S_E S_S) = P(\bar{B}_{UI} \mid S_D S_E S_S) \cdot P(S_D) \cdot P(S_E) \cdot P(S_S) \tag{13.2}$$

where $I = T, C, S$

Example 13.1 The probability of loss of ultimate integrity due to 20% reduction, assuming normal distribution and a $C_V = 0.1$ is as given below.

If this is in relation to compression, then if a 20% reduction has taken place (see Figure 13.1):

$$\Phi\left(\frac{LLR - 0.92\,LLR}{0.92\,LLR \cdot C_V}\right) = \Phi\left(\frac{\dfrac{1}{0.92} - 1}{0.1}\right) = \Phi(0.9) = 1 - \Phi(-0.9) = 0.82$$

$$P(\bar{U}_{UC}) = 0.82 \cdot 0.01 \cdot 0.9 \cdot 0.1 = 0.76 \cdot 10^{-3}$$

Here:

S_D : ultimate damage
S_E : $-65F$
S_S : open hole.

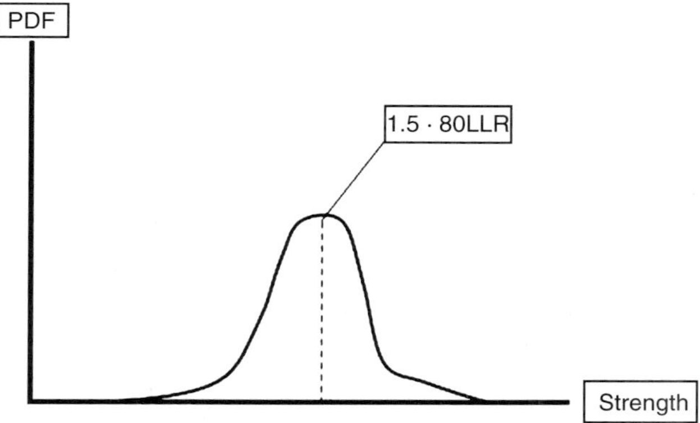

Figure 13.1 20% strength reduction.

The other terms must be compatible.

The same 'states' must be used according to the regulations. No consistent practice currently exists and a firmer hand in establishing criticality is needed for safety management. Process quality standards are specified in the Six-Sigma requirements which must also be part of the regulations, with B-value standard for both panel allowable value and residual strength.

13.5 SAFETY TARGETS

The natural lowest unit for safety level targets is a PSE because, by definition, a failed PSE means a failed airplane, barring special considerations. President Clinton's Commission on Safety of Aviation and Security was chaired in the mid-1990s by Vice-president Al Gore, the report including much status information and many recommendations and views on the future.

The following safety information was included (embellished in terms of 'unsafe states'):

The probability of an unsafe flight is:	10^{-5}
The part caused by structures yields:	10^{-6}
The elements of safety (assuming equal share):	$2 \cdot 10^{-7}$
Equal share per PSE, 100:	$2 \cdot 10^{-9}$
Per DDP depends on model:	>25 variable
Improvement (no increase in the number of fatalities), 10% in 40 years:	$2 \cdot 10^{-11}$

This approach to design requires a model that is consistent with the FE-model, coincides with the DDPs including the locations of defects and damage and sets the level of safety of the PSEs.

13.6 CONCLUSIONS

Targets for structural safety derive from the final report by the White House Commission Safety and Security in Aviation.

The focus is on transport-category aircraft.

The basic unit for safety is the PSE, and the computational unit is the DDP.

Chapter 14
Generalized Safe States

The importance of different integrities to safety has been illustrated in the previous chapters. This chapter describes a generalized version of a 'safe state' based on the initial definition in Backman (2005). The following definition and expansion is the foundation of design requirements and the baseline for updating. The equation of the, "Safe State, 'S_T,' is:

$$P(S_T) = P(S_D S_M S_I S_O S_R) = P(S_D \mid S_M S_I S_O S_R) \cdot P(S_I \mid S_M S_O S_R) \cdot P(S_M \mid S_O S_R)$$
$$\cdot P(S_O \mid S_R) \cdot P(S_R)$$

and the probability of an unsafe state is.

$$P(\bar{S}_T) = P(\bar{S}_D \mid S_M S_I S_O S_R) + P(\bar{S}_I \mid S_M S_O S_R) + P(\bar{S}_M \mid S_O S_R)$$
$$+ P(\bar{S}_O \mid S_R) + P(\bar{S}_R) \tag{14.1}$$

The following events are involved:

S_D : safe design
S_I : safe maintenance
S_M : safe manufacturing
S_O : safe operation
S_R : safe requirements
$\bar{S}_E$: unsafe element.

Two of these events are now defined:

$$\bar{S}_D = S_{DE} \cdot S_{DP}$$

S_{DE} is design execution, S_{DP} design process;

$$S_M = S_{PC} \cdot S_{QC} \cdot S_{ME}$$

S_{PC} is process control, S_{QC} is quality control, S_{ME} manufacturing execution.

Example 14.1 We now evaluate the numerical consequences of a 'best worst-case scenario':

$$P(\bar{S}_D \mid \cdots) \, 10^{-6} \cdot 10^{-3} + 10^{-9} = 2 \cdot 10^{-9}$$
$$P(\bar{S}_M \mid \cdots) \, 3 \cdot 10^{-9}$$
$$P(\bar{S}_I \mid \cdot) = 1.8 \cdot 10^{-9}$$

$$P(\bar{S}_O \mid \cdot) = 10^{-10}$$

$P(\bar{S}_R) =$ too large; process should be changed to 10^{-9}

TOTAL: $\sim 8 \cdot 10^{-9}$

This coincides with the requirement for a structure with 50 PSEs and an average of 25 DDPs, which underscores the advantage of focusing on PSEs to allow for different requirements for different PSEs. It should be noted that the example shows the total effects of all the elements of safety. The numbers are based on a priori probabilities for mechanical damage size and detection.

It is evident that, under these circumstances, the requirements need to be approached with more objectivity and rigor. Present status demonstrates a number of diverse practices and inadequate safety considerations. The process of updating and adding of regulations for composites should be a totally objective process and, as the example illustrates, the quality requirement for a safe state due to defects in the updating process is at least 6σ.

Present practices do not consistently require use of B-values for residual strength, structural panel B-values for buckling, damage included for panel buckling, or allowable values, to mention a few. Safety management and requirements for many difficult scenarios make it necessary to protect both ultimate margin of safety and fail-safe integrity. The demonstration of fail-safe integrity under load has become a necessity in order to be able to demonstrate compliance for fall-back integrity in the presence of both local and widespread manufacturing defects; however, this is only partially included in present regulations, advisory circulars or practices.

Safety management and generalization to the elements of safety extends the number of safety threats to a safe state and the focus measures of monitoring in service are increased from 'undetected loss of damage tolerance integrity' to include 'undetected loss of fail-safe integrity' and 'undetected loss of ultimate integrity'. This expansion is a consequence of defects associated with the total set of elements of safety.

There is also a great deal more pressure on the monitoring process in service to maintain and restore level of safety. The basis for inspection, the foundation of monitoring, also includes failed load-paths, and the detection of hard-to-detect defects that affect structural properties and internal design loads. The first step toward a generalized safety preservation scheme is the understanding of the role of fail-safety.

The safety management monitoring process will require inputs to design in order to deal with hard-to-detect widespread defects, especially those due to process failures that are introduced into service and that must be guarded against or detected. The next section deals with fail-safe integrity and describes the role of ultimate integrity.

Finally the monitoring of service data and inspection results includes different reviews of 'mechanical damage'. It starts with the Bayesian updating of existing probabilities on a PSE basis. These are: probability of damage (marginal distribution of all damages larger than ultimate design damage, D_u), the probability of different damage sizes, and an indirect investigation of 'surviving' defects (including mishaps in manufacturing processes). The latter can be achieved by comparison of probabilities of PSEs between different airplanes and also between different airlines to detect significant deviations in probabilities that could be signs of hard-to-detect, widespread defects. This might be achievable under the auspices of the FAA and other international agencies. The entire process would speed up the accumulation of service experience in a field that is innovative by definition, because of the steady stream of new materials, new processes and new structural concepts.

14.1   FAIL-SAFE INTEGRITY

Fail-safe integrity is fundamental to safety, often because defects can be widespread and difficult to detect in service. The probability of loss of fail-safe integrity can be written as

$$P(\bar{U}_{FS}) = P(\bar{A}_{LP}\bar{R}_R\bar{X}_R\bar{U}_{UR}\bar{H}_D \mid C_L) + P(\bar{A}_{LP}R_R\bar{X}_E\bar{U}_{RD}\bar{H}_D \mid C_L)$$

Here the first term describes a failed redistribution of internal loads, the second a successful redistribution of internal loads, but defects imparted to the participating element before this scenario occurs are present.

These terms are now expanded into the participating events to support the illustration of order of magnitude, starting with $P(\bar{U}_{FS1})$:

$$P\bar{U}_{FS1}) = P(\bar{U}_{RU} \mid \bar{A}_{LP}C_L\bar{R}_R\bar{X}_R\bar{H}_D) \cdot P(\bar{H}_D \mid \bar{A}_{LP}\bar{R}_R\bar{X}_RC_L) \cdot P(\bar{R}_R \mid \bar{A}_{LP}\bar{X}_RC_L)$$
$$\cdot P(\bar{X}_R \mid \bar{A}_{LP}C_L) \cdot P(\bar{A}_{LP} \mid C_L) \tag{14.2}$$

and the second term:

$$P(\bar{U}_{FS1}) = P(\bar{U}_{RU} \mid \bar{A}_{LP}R_R\bar{X}_{RP}\bar{H}_DC_L) \cdot P(\bar{H}_D \mid \bar{A}_{LP}R_R\bar{X}_{RP}C_L) \cdot P(R_R \mid \bar{A}_{LP}\bar{X}_{RP}C_L)$$
$$\cdot P(\bar{X}_{RP} \mid \bar{A}_{LP}C_L) \cdot P(\bar{A}_{LP} \mid C_L) \tag{14.3}$$

These two equations will be used to investigate numerical orders of magnitudes. Participating events are

$\bar{U}_{FS}$: loss of fail-safe integrity
C_L : limit load assumed given in integrity evaluation

R_R : success of internal load distribution

U_{UR} : preserved ultimate Integrity of remaining structure

A_{LP} : focus load path preserved.

Example 14.2 A numerical interpretation of Equations (14.2) and (14.3) is as follows:

$$P(\bar{U}_{FS}) = 0.1 \cdot 10^{-3} \cdot 10^{-2} \cdot 10^{-1} \cdot 10^{-1} + 0.1 \cdot 10^{-3} \cdot 10^{-1} \cdot 10^{-2} \cdot 10^{-2}$$
$$= 10^{-8} + 10^{-9} = 1.1 \cdot 10^{-8}$$

The important conclusion here is that criticality in composite structure is a very important subject, particularly loss of ultimate integrity with its ties to fail-safe integrity and the further question of limit integrity for a number of different states, taking into consideration the different types of integrity.

If an event involves widespread defects caused by one process failure and 100 DDPs are involved, then:

$$P(\bar{U}_{FSP}) = 10^{2} \cdot 1.1 \cdot 10^{-8} \cdot P(\bar{X}_M T_1 V_{13}) = 1.1 \cdot 10^{-6} \cdot 0.34 \cdot 10^{-5} = 3.7 \cdot 10^{-10}$$

which for a fleet of 1000 airplanes yields

$$P(\bar{U}_{FSPF}) = 3.7 \cdot 10^{-7}$$

Again, a realistic assessment requires a specific objective.

14.2   DESCRIPTION OF INTEGRITY

Normal practice in the structural design of composite structure is to consider combinations of states of damage, states of environments (e.g. temperature) and states of local internal, load concentration, e.g. associated with mechanical fasteners. The governing definition of ultimate, internal loads is

$$P(\bar{U}_U) = \sum_{j=1}^{n} \left(\sum_{i=1}^{3} P(\bar{U}_{Uj} \mid S_{Ti}) = \sum_{j=1}^{n} \sum_{i=1}^{3} P(\bar{U}_{Uj} \mid S_D S_{Ei} S_S.) \right) \tag{14.4}$$

Here j represents the different modes of failure considered, e.g. open hole compression, *CAI*, *TAI*, … , filled hole tension, buckling, crippling.

Equation (14.4) can be rewritten as

$$P(\bar{U}_U) = \sum_{j=1}^{n} \left(\sum_{i=1}^{3} P(\bar{B}_{Uj} \mid S_D S_{Ei} S_s) \right) \tag{14.5}$$

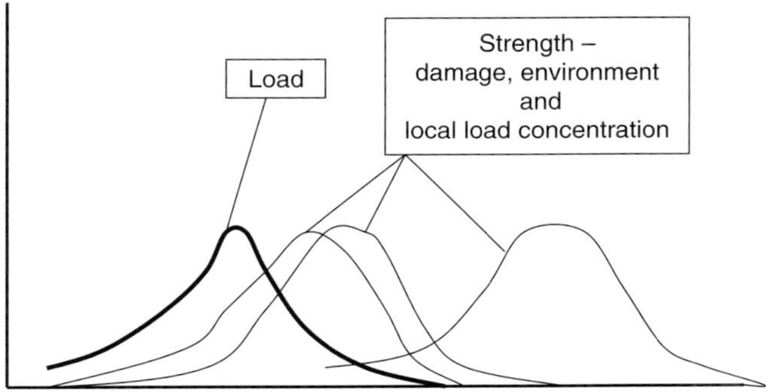

Figure 14.1 Variation in mode and state.

Here:

$\overline{B}_{Uj}$: The event allowable value less than requirement

S_{Ei} : Different states of temperature.

Figure 14.1 shows the typical ultimate design situation covering different modes and gives an internal load definition. Example 14.3 illustrates a set of different strength distribution values with a 'critical' lowest allowable value.

Example 14.3 We assume that the strength distribution furthest to the left is the lowest and is the B-value basis, $0.87\mu_1$. The strength distribution second from the left has a factor of a strength value larger by 1.1.

$$\Phi\left(\frac{ULR - 1.15 \cdot ULR}{1.27C_V ULR}\right) = \Phi\left(-\frac{0.27}{0.1}\right) = \Phi(-2.7) = 0.003$$

The third distribution from the left is larger by a factor of 1.2,

$$\Phi\left(\frac{ULR - 1.2 \cdot 1.15\,ULR}{1.38 \cdot 0.1}\right) = \Phi(-3.8) = 0.00007$$

and the fourth, even larger, is of the order of magnitude of $<10^{-6}$, which yields

$$P(\overline{U}_U) \approx 0.1 + 0.003 + 0.00007 + 10 \cdot 10^{-6}$$
$$= 100 \cdot 10^{-3} + 3 \cdot 10^{-3} + 0.07 \cdot 10^{-3}$$
$$+ 0.01 \cdot 10^{-3} = 103.08 \cdot 10^{-3} \approx 0.1$$

This example makes the point that the critical distribution very quickly dominates the probability of loss of ultimate integrity with this kind of spread of orders of magnitude and number of terms.

The conclusion is that all temporary states enter into the preservation of ultimate integrity even though rare events are involved and although, for example, maximum temperature and limit load must be coincident.

Example 14.4 The relative probability of losing the three kinds of integrity is determined using realistic orders of magnitude, as has been consistently done elsewhere in the examples here in order to meet realistic requirements.

The probability of maximum temperature and limit load coinciding defines the probability of a factor of safety of 1.5. The following equation captures the situation using the same order of magnitude as in the numerical investigation:

$$\Pr\{SF \geqslant 1.5\} = P(S_E C_L) = P(S_E) \cdot P(C_L) = 10^{-3} \cdot 0.33 \cdot 10^{-4} = 3.3 \cdot 10^{-8}$$

If the damage tolerance requirements include T_{MAX}, then the probability of loss of limit integrity is

$$P(\overline{U}_{Li}) = P(S_D S_E C_L) = P(S_D) \cdot P(S_E) \cdot P(C_L) = 10^{-5} \cdot 10^{-3} \cdot 0.33 \cdot 10^{-4} = 3.3 \cdot 10^{-13}$$

Fail-safe integrity requires ultimate integrity of the structure remaining after one load-path has failed in relation to external limit load, and, if T_{MAX} is included, the probability of lost fail-safe integrity is

$$\begin{aligned}
P(\overline{U}_{FS}) &= P(\overline{A}_{LP}\overline{R}_{IL}S_E C_L) + P(\overline{A}_{LP}R_{IL}\overline{B}_{RS}S_E C_L) \\
&= 10^{-2} \cdot 10^{-3} \cdot 10^{-1} \cdot 10^{-3} \cdot 0.33 \cdot 10^{-4} + 10^{-2} \cdot 10^{-3} \cdot 0.9 \cdot 0.2 \cdot 10^{-3} \cdot 0.33 \cdot 10^{-4} \\
&= 3.3 \cdot 10^{-11} + 0.6 \cdot 10^{-10} = 0.93 \cdot 10^{-13}
\end{aligned}$$

Here the following events are included:

$\overline{A}_{LP}$: failure of focus load path
$\overline{R}_{IL}$: failure to redistribute internal loads
S_E : maximum temperature
$\overline{B}_{RS}$: strength is less than requirement.

The result shows that the probability of losing ultimate integrity is larger than the probability of losing damage tolerance integrity or fail-safe integrity. Allowable B-values are used for design.

14.3 SAFETY MANAGEMENT STRATEGY

Widespread defects owing to autoclave process failure (or equivalent) present a very difficult situation for composite structure. When the resulting defects involve reduction of structural properties, the use of design criteria to create a manageable situation seems to be an effective approach to take. For example, if ultimate strength properties are reduced, fail-safe integrity will be lost as this type of defect is consistently hard to detect. The use of fail-safety as a fall-back position to damage tolerance integrity would not be a viable approach. However, if design criteria (specifically the limit load requirements are prescribed, such that damage tolerance becomes a critical requirement for design) require positive ultimate margins of safety with a matched specific value compatible with the quality process requirements (e.g. 6σ quality could include up to 20% reduction), fail-safe integrity would be preserved, and would protect a fail-safe fall-back. If damage tolerance residual strength, for example, is reduced at the same time, not an uncommon situation, the fall-back of fail-safe detail design would be available.

Structural design would create a safe situation with conventionally problematic events. A sound set of design criteria could take care of this very difficult type of defect and create a foundation for fail-safety.

It is a fact that composite structure is in a perpetual state of innovation. New materials, new processes and new structural concepts pass by in a steady parade. Service experience is slow to accumulate and changing data create an environment of uncertainty. Therefore a process for monitoring, analysis, risk management and mitigation of uncertainty is needed. This is, as is the case with biological processes, a time to turn weakness into strength. Mechanical damage detection can be used to update initial probability distributions (Bayesian updating) for use in risk management and uncertainty reduction.

One can look at the service monitoring and updating of probabilities as a part of risk management based on the probability of an unsafe state, starting with the input from manufacturing and with the assumption that all processes start with a 6σ quality.

All unacceptable defects lie outside the 6σ interval, and manufacturing, among other areas, produces defects with the following effects:

1. reduced structural properties;
2. increased internal loads;
3. spurious mechanical damage;
4. chemical and physical sensitivity (long-term degradation).

The probability of these defects can be written as

$$P(\bar{X}_M T_i V_{i3}) = P(V_{i3} \mid T_i \bar{X}_M) \cdot P(T_i \mid \bar{X}_M) \cdot P(\bar{X}_M) \tag{14.6}$$

The following events are involved:

$\overline{X}_M$: failed manufacturing process
T_i : Type of defects
V_{i3} : Extent of defect.

Both reduced structural properties and increased internal loads represent effects of defects that can be present from roll-out and constitute a basic safety management defect. Safety monitoring is discussed in the next section together with updating, risk management and uncertainty reduction.

We have previously discussed monitoring and updating probabilities of damage, sizes, and detection requirements. We have touched on uncertainty of damage, residual strength and detection. It is evident that service monitoring is a very important process.

14.4   UNCERTAINTY AND MONITORING

The basic quantity involved in risk management is the probability of an unsafe state. It can be written as

$$P(\overline{S}_T) = P(\overline{X}_t \overline{H}_t \overline{B}_{LT} \overline{X}_{LT} D_{5T} \overline{H}_T) = P(\overline{H}_t \mid \overline{X}_t) \cdot P(\overline{B}_{LT} \mid D_{5T} \overline{X}_{LT} \overline{H}_T)$$
$$\cdot P(\overline{H}_T \mid \overline{X}_{LT} D_{5T}) \cdot P(D_{5T} \mid \overline{X}_{LT}) \cdot P(\overline{X}_{LT}) \tag{14.7}$$

The factors on the right-hand side are as follows:

The first factor is the probability of detecting damage in a marginal sense – a quality value.

The second is the probability of limit residual strength less than limit requirement, LLR.

The third probability is for the detection of D_5.

The fourth factor is the probability of size region D_5.

The fifth factor is the probability of large-scale damage in general.

Many uncertainties and updating requirements are involved.

Figure 14.2 illustrates the uncertainty of detection of damage in composite structure.

Another type of uncertainty is described in Figure 14.3.

Both severity and external damage must be part of determination, if significant.

The risk management carried out will be based initially on damage size distribution and probabilities of large damages. The starting value for equation can be selected from the results in the example following.

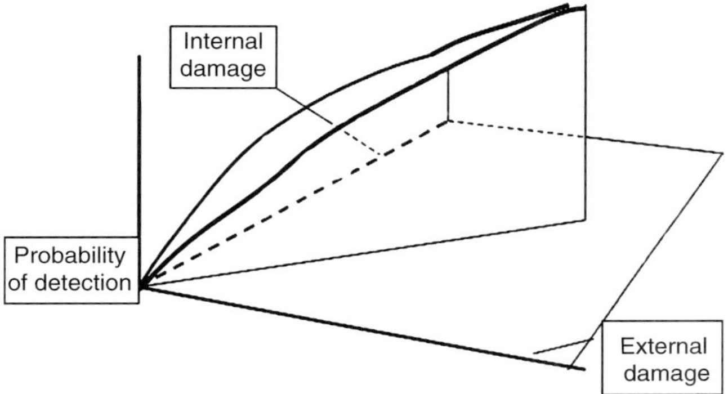

Figure 14.2 Detection and internal and external damage.

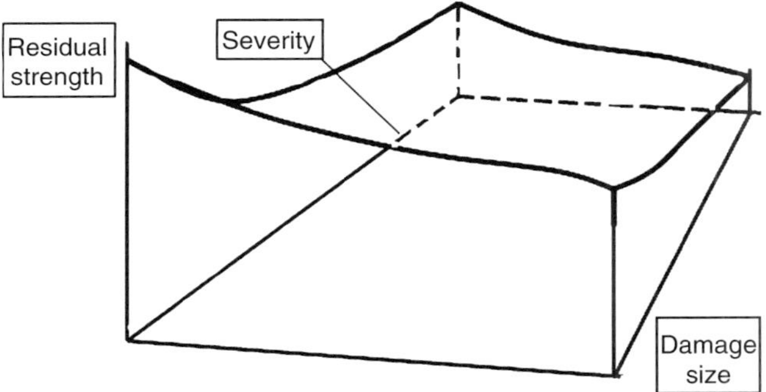

Figure 14.3 Residual strength damage and severity.

Example 14.5 This example will be used to demonstrate risk management. Equation (14.7) contains marginal probability of detection, the probability of limit damage sizes and probability of 'large damage'. The Bayesian approach has produced updated values; the resulting probability of an unsafe state, $>2 \cdot 10^{-11}$, needs attention. The probability of survival for the design mission in number of flights is

n flights	Survival of n flights	Fail-safe survival
1	–	–
100	0.98	–
1000	0.79	–
2000	0.63	Inspection
2500	0.56	–

(Continued)

n flights	Survival of n flights	Fail-safe survival
3000	0.50	–
If not, load path fails		Survival!
–	–	–
10	0.79	10
20	0.59	20
Damage must be found		
30	0.45	30
40	0.35	40

The risk management process concludes, for example: new inspection before 2500 flights, otherwise fail-safe could be activated.

The safety-conscious approach places trust in risk management, but both detection reliability and residual strength prediction must concomitantly work in parallel to damage probability.

14.5 CONCLUSIONS

Risk management and uncertainty mitigation are important parts of design, manufacturing and service.

Monitoring and service databases are important aspects of all cradle-to-grave activities, improving both education and aviation.

The definitions of the realistic, internal, loads design mission are vital parts of safety management.

Chapter 15
Scenarios and Safety

Many different types of situation arise in relation to safety management and elements of safety, along with many combinations of criticality and difficulty. The value of the numerical investigation of the complications that are possible is hard to overvalue. We will now embark on a review of scenarios where Six-Sigma (6σ) quality could prove very constructive.

15.1 MANUFACTURING PROCESS DEFECTS

Manufacturing processing mishaps that lead to loss of structural properties give rise to many complications. They can often be widespread and affect different properties in a different way.

When ultimate properties are involved we often find that ultimate compression strength, compression stiffness and compression toughness (G_{Ic}) are reduced in similar ways. We have found that the ultimate shear strength, shear stiffness and shear toughness change as a group in the same way as regards time and amount, and finally that ultimate, tension properties are the most insensitive to defects in processing. We have also repeatedly concluded that effects vary in degree from case to case.

With a reduction in tension strength, compression strength and shear strength vary, the same being true for applied internal loads. It is a complicated process to determine what constitutes criticality, but it is undeniable that loss of ultimate integrity leads to a lack of fail-safety, especially when defects are widespread. It is, therefore, as mentioned previously, necessary to provide ultimate integrity with a positive ultimate margin of safety compatible with maximum 'reduction of structural properties'. As has also been pointed out, damage tolerance design provides that opportunity. A prudent selection of limit load damage sizes can for example produce a 15% structural property reduction buffer, and a 6σ quality could include that defect. Figure 15.1 illustrates this situation.

Figure 15.1 shows the region of unacceptability, which in this case could contain defects causing property reduction of greater than 15% and loss of both ultimate integrity and fail-safe integrity. The probability of loss of fail-safe integrity is

$$P(\bar{U}_{FS}) = P(\bar{X}_M T_1 V_{13}) = P(V_{13} \mid T_1 \bar{X}_M) \cdot P(T_1 \mid \bar{X}_M) \cdot P(\bar{X}_M)$$
$$= P(V_{13} \mid T_1 \bar{X}_M) \cdot P(T_1 \mid \bar{X}_M) \cdot \Phi(-6 \cdot \sigma) \tag{15.1}$$

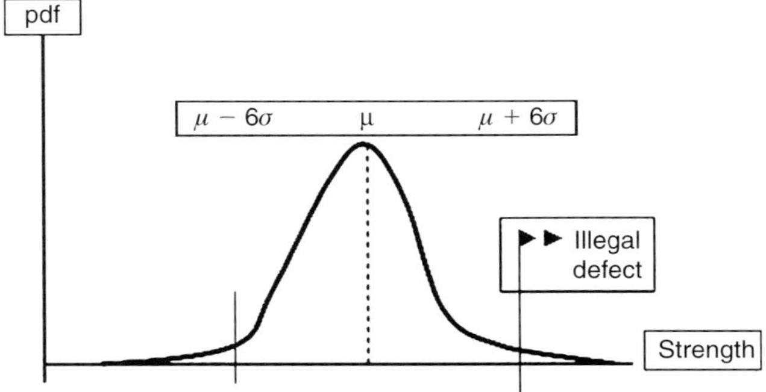

Figure 15.1 Six-Sigma region.

The following events are involved:

V_{13}　　　　: the value of the defect of type 1, in this case maximum, 3
T_1　　　　　: type 1, reduction of ultimate properties
$\Phi(-6\sigma)$　: part of the error function that exceeds 6σ.

The next example presents a numerical illustration of Equation (15.1), giving orders of magnitude for an extreme case.

Example 15.1　　Given here is the 'maximum case' for the 'ultimate defects case'.

$$P(\bar{U}_{FS}) = 10^{-2} \cdot 0.4 \cdot 10^{-9} = 0.4 \cdot 10^{-11}$$

This order of magnitude also creates problems in relation to the 'widespread defects case'; a number of locations, *n*, are involved.

We now focus on the zone involved (*n* adjacent DDPs). The situation (the probability of a zone with preserved ultimate integrity without special design action) can be written as

$$P(U_{ZU} \mid W) = P(U_{ZU1} \ldots U_{ZUn} \mid W) = P(U_{ZU1} \mid U_{ZU2} \ldots U_{ZUn}W)$$
$$\ldots P(U_{ZU(n-1)} \mid U_{ZUn}W) \cdot P(U_{ZUn} \mid W) \Rightarrow$$
$$P(U_{ZU}) = [1 - P(\bar{U}_{ZU1} \mid U_{ZU2} \ldots U_{ZUn}W)] \ldots [1 - P(\bar{U}_{ZU(n-1)} \mid U_{ZUn}W)]$$
$$\cdot [1 - P(\bar{U}_{ZUn} \mid W)] \Rightarrow$$
$$P(\bar{U}_{ZU}) = P(\bar{U}_{ZUn} \mid W)$$

15.1.1 Widespread defects

The above example leads to the conclusion, in relation to widespread defects, that the probability of loss of integrity of a PSE is equal to the probability of loss of integrity of one member in the 'widespread' region.

This clearly supports the need to have a built-in positive ultimate margin of safety that is large enough to protect the PSE from loss of fail-safe integrity. If the damage tolerance property reduction coincides with the reduction of ultimate properties, then the fall-back solution has to be fail-safe design.

The probability of an unsafe state, the event that includes both processing defects affecting residual strength and large damage (threatening damage tolerance integrity), is

$$
\begin{aligned}
P(\bar{S}) &= P(\bar{X}_M T_2 V_{23} \bar{B}_{L0} \bar{B}_{LT} \bar{X}_T D_{5T} \bar{H}_T) \\
&= P(\bar{B}_{LT} \mid \bar{X}_T D_{5T} \bar{B}_{L0} \bar{X}_M T_2 V_{23} \bar{H}_T) \cdot P(\bar{H}_T \mid \bar{X}_T D_{5T}) \cdot P(\bar{B}_{L0} \mid \bar{X}_M T_2 V_{23}) \\
&\quad \cdot P(D_{5T} \mid \bar{X}_T) P(\bar{X}_T) \cdot P(V_{23} \mid \bar{X}_M T_2) \cdot P(T_2 \mid \bar{X}_M) \cdot P(\bar{X}_M)
\end{aligned}
\tag{15.2}
$$

The factors on the right-hand side are as follows:

- The first, the probability of residual strength being less than requirement with manufacturing defects present and mechanical damage in the limit range;
- The second, the probability of mechanical damage not being detected;
- The third, the probability of residual strength being less than requirement with manufacturing defects present;
- The fourth, the probability of limit damage being present;
- The fifth, the probability of damage;
- The sixth, the probability of maximum value;
- The seventh, the probability of a residual strength reduction due to defect;
- The eighth, the probability of a manufacturing defect.

A numerical illustration of this case with Six-Sigma quality follows in the next example:

Example 15.2 This example provides a numerical description Equation (15.2):

$$
P(\bar{S}_T) = 0.8 \cdot 10^{-3} \cdot 0.0025 \cdot 10^{-3} \cdot 10^{-2} \cdot 0.3 \cdot 10^{-9} = 0.06 \cdot 10^{-19}
$$

Here the same requirements are used to describe the probability numbers as in the entire family of examples.

This illustrates that this is a very improbable case when combined with Six-Sigma quality in manufacturing if limit mechanical damage is also inflicted.

The next section will focus on activities related to damage tolerance. The prime concern is that they constitute threats to fail-safety.

15.1.2 *Increase in growth rates and reduction in damage resistance*

Quality control and major inspections play a very important role in identifying: excessive growth rates and reduced damage resistance (quality control, in preventing difficult defects from entering service; and major inspections, in cooperation with monitoring procedures by identifying accumulation of abnormal damage). Thus, while there are many criteria in design that are claimed to be based on the 'no-growth' approach, the actual case falls under the heading of 'proving a negative', which clearly can be invalidated by manufacturing defects, environmentally aggressive situations or damage sizes conducive to growth. It seems that a rational design criterion requires determining maximum growth rates and circumstances. This makes it possible in the monitoring process to evaluate whether anomalous conditions obtain or whether there are regular conditions promoting growth and adding to the requirements for detection.

Damage resistance is a very important property for composite structure. It makes fail-safe a possible fall-back when widespread defects are present and also requires a demonstration of recovery under load (limit load). The combination of reduction of damage tolerance and damage resistance is a scenario fraught with difficulty and not only presents a challenge to fail-safe design, but constitutes a threat to 'Discrete source damage'.

Emerging new requirements involving hailstone strikes in flight also have to be dealt with in terms of meeting new kinds of compliance. It is very important to understand the nature of these kinds of event. The screening of 'new' materials must include determination of sensitivity to loss of damage resistance. The expectation is that defects will arise in both manufacturing and repairing. The probability of reduced damage resistance due to manufacturing defects can be expressed as

$$P(\bar{D}_R) = P(\bar{X}_M T_{32} V_{33} \bar{C}_L \bar{Y}_t \bar{R}_D) = P(\bar{R}_D \mid \bar{Y}_t \bar{C}_L \bar{X}_M T_{32} V_{33}) \cdot P(\bar{Y}_t \mid \bar{C}_L \bar{X}_M T_{32} V_{33})$$
$$\cdot P(\bar{C}_L \mid \bar{X}_M T_{32} V_{33}) \cdot P(V_{33} \mid T_{32} \bar{X}_M) \cdot P(T_{33} \mid \bar{X}_M) \cdot P(\bar{X}_M) \qquad (15.3)$$

The following events are included:

$\bar{X}_M$: manufacturing defect present
T_{32} : defect type is reduced damage resistance
V_{33} : maximum value of defect
$\bar{C}_L$: limit load exceeded
$\bar{Y}_t$: focus load path failed
$\bar{R}_D$: internal load redistribution failed (unsuccessful fail-safe).

Example 15.3 A numerical requirement is now investigated in relation to inadequate damage resistance, $\bar{D}_R$, by the use of Equation (15.3):

$$P(\bar{D}_R) = 0.1 \cdot 0.1 \cdot 0.33 \cdot 10^{-4} \cdot 0.2 \cdot 0.4 \cdot 10^{-9} = 0.33 \cdot 10^{-16}$$

As the result indicates, a rare event is involved. The specifics for the detection for the material system in question in quality control, or the interactions between process control and quality control, are expected to confirm this. If the manufacturing processes are also complemented with a quality control process, we could write:

$$P(\bar{D}_{R0}) = P(\bar{X}_M T_{32} V_{33} \bar{H}_Q) = P(\bar{H}_Q \mid \bar{X}_M T_{32} V_{33}) \cdot P(V_{33} \mid T_{32} \bar{X}_M)$$
$$\cdot P(T_{32} \mid \bar{X}_M) \cdot P(\bar{X}) \tag{15.4}$$

The required probability of not detecting the defect in damage resistance can be deduced from the required order of magnitude of

$$P(\bar{D}_R) = P(\bar{H}_Q \mid \bar{X}_M T_{32} V_{33}) \cdot 0.2 \cdot 0.4 \cdot 10^{-9} \leq 2 \cdot 10^{-11}$$
$$\Rightarrow P(\bar{H}_Q \mid \bar{X} T_{32} V_{33}) \leq 1.25 \cdot 10^{-1}$$

This value corresponds to an attainable probability of detection of ≈ 0.9.

15.2 INCREASE IN INTERNAL LOADS DUE TO MANUFACTURING DEFECTS

The approach of letting damage tolerance cover up to 20% of the ultimate margin of safety will allow an up to 20% increase in internal loads due to variation in bolt-hole sizes or incorrect grip lengths or a lack of clamp-up. The combination of processing defects and problems in fastener installations (including clamp-up) is an improbable event:

$$P(\bar{M}_{M1} \bar{M}_{M4}) = P(\bar{M}_{M1}) \cdot P(\bar{M}_{M4}) = P(V_{13} \mid T_1 \bar{X}_{M1}) \cdot P(T_1 \mid \bar{X}_{M1}) \cdot P(\bar{X}_{M1})$$
$$\cdot P(V_{41} \mid T_4 \bar{X}_{M4}) \cdot P(T_4 \mid \bar{X}_{M4}) \cdot P(\bar{X}_{M4}) \tag{15.5}$$

The coincidence of process defects and fastener installations has this probability based on 6σ requirements:

$$P(\bar{M}_{M1} \bar{M}_{M\$}) = 0.2 \cdot 0.4 \cdot 10^{-9} \cdot 0.2 \cdot 0.3 \cdot 10^{-9} = 0.5 \cdot 10^{-20}$$

This describes a very improbable event that can be extended to a general conclusion about independent processes, in this context. It makes it possible to use the positive ultimate margin of safety due to the criticality of damage tolerance discussed in previous sections.

Scheduled regular maintenance should be conducted according to a 6σ process with maximum defects limited to an effect of a 15% load increase, subject to regular inspections. If a structural function is involved (e.g. jack-screw), adequate safety would be provided if the ultimate margin of safety guaranteed by the damage tolerance criticality and special damage sizes apply for both elements and attachments.

15.3 MAINTENANCE DEFECTS

Maintenance is a very important element of safety and often has solutions very similar to those for the safety challenge that applies to manufacturing, as seen above.

The structural repair aspects of maintenance also have requirements very similar to those to manufacturing and, in addition, unique difficulties with process control (especially for bonded repairs). In addition to meticulous process control, it seems that structural repairs should be designed to be fail-safe and examined at the regular structural inspections.

The regular inspection cycles need to be coordinated with the monitoring process. The specific objectives include follow-up of anomalous results in terms of frequency and density. The requirement is that load-path failures be identified separately for repairs and PSEs, and that coordination of changes in regular inspections is effected in order to satisfy changing safety requirements due to updating.

A prudent organization of maintenance draws the following conclusions in relation to implementation of Six-Sigma processes:

- One oversight in implementing scheduled maintenance is a defect out with the process.
- One 'missed' step in the requirements for implementation represents a defect out with the process.
- A comparison with previous repair records for the same PSE is required for at least one other airplane of the same model.

The resulting probability for the first two defects listed above is 10^{-9}, the third requirement assuring up-to-date damage probabilities.

15.4 DEFECTS IN OPERATION

The following controlled processes should be part of operation:

- Do not exceed limit load.
- Avoid extreme environments and weather (turbulence, hailstorms, thunderstorms, etc).

- Conduct walk-around inspections and report results.
- Supervise ground operation and report mishaps.

These four sub-elements of operation should satisfy a quality of Six-Sigma, which would guarantee a probability of less than 10^{-9} for each process.

15.5 DEFECTS IN REQUIREMENTS FORMULATION

The FAR 25 regulations constitute the backbone of structural requirements. These are minimum requirements and presently do not contain any explicit to safety. They do, however, contain provisos such as 'Do so, provided not impractical', which clearly sends out the message that one does not have to satisfy safety requirements, if this proves to be impractical. Engineering solutions should have realistic safety solutions as a minimum.

This element of safety has a very specific effect and must have a clearly understandable safety background, because it has a definite impact on practices both existing and evolving. The present situation is that there are no existing regulations for composite airplane structure. Current regulations need specific allowable values definitions and requirements that should continue to expect fail-safe detail design for their use. The solution to widespread processing defects for B-value designed structure is problematic and consequently fail-safe design requirements are not satisfied.

Thus the definitions of ultimate allowable values are important not only for ultimate factor of safety, but are needed to assure fail-safety. The consistency of allowable values should be enforced. Present practices of, for example, compression allowable B-values, contain an assortment of practices for 'open-hole' compression, compression after impact, buckling allowable values, crippling allowable values and damage tolerance allowable values that are all competing for criticality and, depending on location in the aircraft, are defined by:

- a state of damage, which can be ultimate damage, limit damage and fail-safe damage;
- a state of environment, which for example can involve maximum temperature, RT or $-65F$;
- a state of disturbance, which for example can be open-hole or filled-hole.

The minimum concern would require the use of B-values and either the critical situation, or that which prevails at the actual location. Present practices should be brought to a state that favors consistency, in particular a change of the mode of failure in the analysis that does not alter the state of damage and the state of the environment.

The present definition for design (in DFR 25) 'Limit load is the largest load expected in service', does not constitute a useful one. For example, it can be asked: What load? Whose expectation? How many times? What probability? What load cases?

A definition of limit load for design should be based on internal loads and obtaining environmental states combined with state of damage at the pertinent location and the probability of detection at the same location. In short a realistic definition of limit load, damage size and environment, which all are important to composite, structural design. Special requirements for dynamic loads, e.g. 'stall buffet', must be included. The definition of maximum temperature, typical temperature and required local temperature procedures and compliance demonstrations are imperative, and are particularly important for hybrid structures.

This section could continue almost indefinitely, but suffice it to say that regulations specifically applicable to composite structures are urgently needed and should be formulated and documented by the technical experts not the bureaucrats.

The importance of loss of limit integrity for composites and the safety requirement of B-value residual strength allowable values, makes it particularly important to specify a 'mission description' that identifies the internal load distribution for local structure in association with lost damage tolerance integrity.

The foundation of transport-category safety requires modern, specific regulations and compliance requirements for composite structure. Claims that the regulations for metal aircraft are applicable to composite structure are flimsy. Very little carries over to composites and much is missing. In addition, The 'composites world' is not a homogeneous world. A steady stream of new materials, new processes and new structural concepts have passed by in almost endless review, resulting in very slow acquisition of experience and service databases.

The two elements of safety that are currently of particular importance are manufacturing and requirements formulation, each of which is struggling with the consequences of steady change.

The five elements of safety discussed in this chapter and others have demonstrated that the nature of the quest for safe composite structures requires an organized identification of similarities and differences between material systems and material processes in the service databases. In these times of 'centers of excellence' it would appear very productive to establish an industry-sponsored educational arm. The practical elements of composite design are hopelessly neglected in academia, and the claims of applicability of metal material behavior in the regulations do not help much.

The different strategy of composite structural design discussed in the last two chapters is just a beginning of the influence of safety-based design constraints. However, the advances beginning to materialize may very well make it possible to introduce safety improvements of an order of magnitude sufficient to meet the requirements in relation to accident rates proposed by the Commission of Safety and Security in Aviation.

15.6 CONCLUSIONS

A practical 6σ process with a definition of no defects within the 6σ region has a manageable order of magnitude if a gradual increase occurs to between 6σ and 7σ from a 'zero defect' to a defect with a 10% effect, and provides slow growth moderation.

The ultimate allowable value used in design and criticality evaluation need to be based on a consistent definition.

Limit load definitions in the design context should be based on internal loads.

A data-based intervention for different kinds of load (tension, compression and shear), and for mechanical fasteners (both permanent and for repair), should be used in design.

Reliable design and requirements formulation processes are necessary to ensure safe and consistent designs promotion and to produce useful service data.

Chapter 16
Design Mission and Philosophy

The consequences of the design strategy associated with widespread manufacturing defects and the frequency of major inspections require the use of updated data on probability of damage in order to change the time to the 'next' major inspection. The major challenge is the need to preserve the use of fail-safe integrity by using ultimate margin of safety back-up to protect against 15% ultimate strength loss in association with the control of damage tolerance criticality.

16.1 MANEUVER CRITICAL DESIGNS

The time between loss of limit integrity and failure of a principal structural element (PSE) is the critical variable. We will assume normally distributed variables for both strength and internal loads. We assume that the requirements lead to B-values for residual strength. Figure 16.1 describes the strength distribution.

The probability requirement of an unsafe state is the major design criterion and can be written in its simplest form as

$$P(\bar{S}_T) = P(\bar{H}_\tau \bar{U}_T \bar{H}_T) = P(\bar{H}_\tau) \cdot P(\bar{H}_T \mid \bar{U}_T) \cdot P(\bar{U}_T)$$

$$= P(\bar{H}_\tau) \cdot P(\bar{H}_T \mid \bar{X}_T D_5) \cdot P(\bar{B}_L \mid \bar{X}_T D_{5T}) \cdot P(D_{5T} \mid \bar{X}_T) \cdot P(\bar{X}_T) \leq$$

$\leq$ which depends on the number of PSEs and DDPs in the design *model*. A realistic order of magnitude is $\leq 10^{-11}$ which could for example result in

$$P(\bar{S}_T) \leq 10^{-2} \cdot 10^{-3} \cdot 10^{-1} \cdot 10^{-3} \cdot 10^{-2} = 10^{-11}$$

The important values are the detection of damage in size region D_5 given damage, B-value for residual strength allowable values and probability of damage in the D_5 damage size given that damage is present.

The distribution of limit, internal loads is based on critical damage tolerance and 15% ultimate positive margin of safety. The loads are assumed to be maneuver loads and the bulk of the distribution is centered on

$$N = \frac{1.2 \cdot 1.5}{2.5} LLR = 0.7 LLR \Rightarrow \mu_y = 0.7 LLR$$

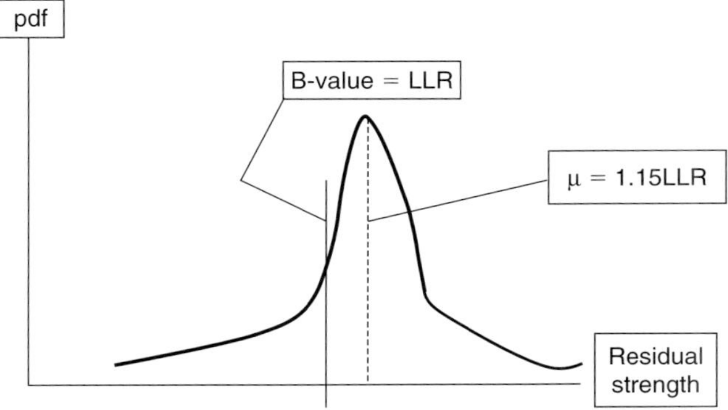

Figure 16.1 Strength distribution.

Here the distribution parameter represents a 'well-controlled' flight except for a few limit load excursions. It is interesting to see that the mean emphasizes the need for discrete source damage design and compliance demonstration of damage resistance.

Strength distribution has $\mu_x = 1.15LLR$

The resulting probability of failure density function has the following parameters:

$$\mu_g = (1.15 - 0.7)LLR = 0.45LLR$$

and

$$\sigma_g = (1.32 + 0.49)^5 0.1 \cdot LLR = 0.13LLR$$

and

$$P(\bar{A}) = \Phi\left(-\frac{0.45}{0.13}\right) = 0.00041$$

and

$$P(A) = 0.99959$$

The probability of failure after 'n' flights:

n flights	Probability of survival for n flights	Probability of failure before or at the nth flight
1	0.99959	0.00041
100	0.96	0.04
1000	0.66	0.34
3000	0.29	0.71
6000	0.09	0.01

If the back-up fail-safety is activated the following results apply (one load path is failed):

	Probability of survival
n flights	for n flights
1	0.735
10	0.05

The failure of a load-path obviously must be uppermost in mind during post-flight inspections.

Monitoring changes in the probability distributions is of importance for achieving safety in the inspection period. The importance of having a fail-safe fall-back is shown in the second table. The probability of this survival value is based on a fail-safe integrity with 15% reduction of ultimate strength, which again demonstrates the importance of damage resistance during internal loads redistribution after single load-path failure.

As discussed previously, the answer to a loss of fail-safe integrity for a Six-Sigma quality with a legal defect level of 15% ultimate strength reduction is a damage tolerance criticality that gives ultimate design a 15% margin of safety. The probability of a 15% ultimate strength reduction is

$$P(\overline{S}_{M0}) = 10^{-1} \cdot 0.5 \cdot 10^{-9} = 5 \cdot 10^{-11}$$

This obviously is only acceptable when the damage tolerance requirements are 15% more critical than ultimate strength, and will not satisfy the increased safety level of one order of magnitude proposed by the FAA.

16.2 GUST CRITICAL STRUCTURE

If horizontal stabilizer is treated separately we find that the incremental gust load factor is less than $\Delta n_z < 1.2$, and the following mean value approximately applies:

$$\mu_y = 0.8LLR$$

yielding somewhat more favorable results than do maneuver critical structures. A demonstration of rigorous compliance in relation to safe periods for survival after loss of limit integrity is clearly required. The type of criticality is very important in the internal loads probability distribution when determining probability of survival.

16.3 WIDESPREAD DEFECTS

The case of widespread defects, e.g. reduction in ultimate strength, is difficult and would be best handled by a process that would include quality control to satisfy the following equation:

$$P(\overline{W}_S H_{QC}) = P(\overline{X}_M V_{13} T_1 H_{QC}) = P(H_{QC} \mid T_1 V_{13} \overline{X}_M) \cdot P(V_{13} \mid T_1 \overline{X}_M)$$
$$\cdot P(T_1 \mid \overline{X}_M) \cdot P(\overline{X}_M)$$

and the numerical illustration:

$$P(\overline{W}_S H_{QC}) = 10^{-3} \cdot 10^{-1} \cdot 0.5 \cdot 10^{-9} = 5 \cdot 10^{-14}$$

this result would eliminate wide-spread damage as a concern if combined with damage monitoring and comparison.

16.4 STRUCTURAL INTEGRITY AND COMPOSITES

The evolution of structural integrity in terms of practices in the 'composites world' has brought about change of concepts and contents respectively. Damage tolerance integrity, discrete source integrity, fail-safe integrity and damage resistance integrity have through practice, evolved in to something more complicated than that which existed previously in the subsonic, 'metal world'.

The first change involved the evolution of limit and ultimate allowable values and their relation to integrity. Damage tolerance integrity and ultimate integrity have evolved to yield an expression of the following type:

$$P(\overline{U}_U) = P(\overline{B}_U S_D S_E S_S S_{UL}) = P(\overline{B}_U \mid S_D S_E S_S S_{UL}) \cdot P(S_{UL} \mid S_D S_E S_S) \cdot P(S_D)$$
$$\cdot P(S_E) \cdot P(S_S) \cdot P(S_{UL}) \tag{16.1}$$

The sub-events are:

$\overline{U}_U$: loss of ultimate integrity
$\overline{B}_U$: strength is less than ultimate requirements USR
S_D : state of damage
S_E : state of environment, e.g. maximum temperature
S_S : local state of response, disturbance e.g. open hole
S_{UL} : ultimate state of internal load.

The implication is that these events or states are concurrent.

The total structural integrity, U_{SI}, can be written for composite structure as a mixture of old and new as:

$$P(U_{SI}) = P(U_U U_L U_{FS} U_{DS} U_{DR}) = P(U_U \mid U_L U_{FS} U_{DS} U_{DR}) \cdot P(U_L \mid U_{FS} U_{DS} U_{DR})$$
$$\cdot P(U_{FS} \mid U_{DS} U_{DR}) \cdot P(U_{DS} \mid D_{DR}) \cdot P(D_{DR})$$

$$(16.2)$$

The sub-events are:

$$
\begin{array}{ll}
U_{SI} & : \text{Preserved structural integrity} \\
U_U & : \text{Preserved ultimate integrity} \\
U_L & : \text{Preserved limit integrity} \\
U_{FS} & : \text{Preserved fail-safe integrity} \\
U_{DS} & : \text{Preserved discrete source integrity} \\
U_{DS} & : \text{Preserved damage resistance integrity.}
\end{array}
$$

The loss of structural integrity, $\bar{U}_{SI}$, can be written in the following approximate manner, making it possible to identify the influence of all elements of safety, in terms of:

- reduced structural properties
- limit load exceedances
- spurious mechanical damage
- misleading regulations and requirements processes.

The probability of loss of structural integrity is

$$P(\bar{U}_{SI}) = P(\bar{U}_U \mid U_L U_{FS} U_{DS} U_{DR}) + P(\bar{U}_L \mid U_{FS} U_{DS} U_{DR})$$
$$+ P(\bar{U}_{FS} \mid U_{DS} U_{DR}) + P(\bar{U}_{DS} \mid U_{DR}) + P(\bar{U}_{DR}) \qquad (16.3)$$

The terms of the right-hand side of Equation (16.3) are:

The first term: loss of ultimate integrity while the other four are inviolate;
The second term: loss of limit integrity while fail-safe, discrete source and damage resistance integrities are intact;
The third term: loss of fail-safe integrity with intact discrete source and damage resistance integrities;
The fourth term: loss of discrete source integrity and intact damage resistance integrity;
The fifth term: loss of damage resistance integrity.

This formulation makes it possible to include effects of defects and damage due to all identified elements of safety and a generalized version of the probability of an unsafe state.

$$P(\bar{S}_{SI}) = P(\bar{U}_{SI} \cdot (\bar{H}_T \bar{X}_t \cup \bar{H}_{QC}) \cdot (\bar{X}_M \cup \bar{X}_I \cup \bar{X}_O \cup \bar{X}_R)) \qquad (16.4)$$

Equation (16.4) describes the probability of an unsafe state as the sum of probabilities of:

- lost structural integrity with undetected mechanical damage;
- lost structural integrity with undetected, unacceptable manufacturing defect;
- lost structural integrity with undetected, unacceptable maintenance defect;
- lost structural integrity with undetected, unacceptable operation defect;
- lost structural integrity with undetected, unacceptable requirements.

The concise version of 'lost structural integrity' is expressed as

$$P(\bar{U}_{SI}) = \sum_{k=1}^{5} P(\bar{U}_{SI})_k \qquad (16.5)$$

which is used in the numerical illustration below:

$$P(\bar{S}_{SI})_1 \quad : \quad P(\bar{U}_{SI})_1 \cdot P(\bar{H}_T \mid \bar{X}_t) \cdot P(\bar{X}_t) + O(10^{-16}) = 10^{-6} \cdot 10^{-3} \cdot 10^{-2} = 10^{-11}$$

$$P(\bar{S}_{SI})_2 \quad : \quad 0.02 \cdot 10^{-2} \cdot 10^{-9} + 0.68 \cdot 10^{-2} \cdot 10^{-9} + O(10^{-16}) = 0.7 \cdot 10^{-11}$$

$$P(\bar{S}_{SI})_3 \quad : \quad 0.02 \cdot 10^{-11} + 0.68 \cdot 10^{-11} + O(10^{-16}) = 0.7 \cdot 10^{-11}$$

$$P(\bar{S}_{SI})_4 \quad : \quad 0.68 \cdot 10^{-11}$$

$$P(\bar{S}_{SI})_5 \quad : \quad 0.4 \cdot 10^{-11} + 0.4 \cdot 10^{-11} + 0.4 \cdot 10^{-11} = 1.2 \cdot 10^{-11}$$

$$\sum P(\bar{S}_{SI})_k \quad : \quad 4.3 \cdot 10^{-11}$$

This numerical illustration represents a compromise between either improving quality control by two orders of magnitude or selecting the limit damage sizes so that limit becomes 20% more critical than ultimate, and thereby solves the 'widespread' defects problem. A comparison in this example which is based on Six-Sigma quality of the participating processes shows that an initial, equal share of 'safety responsibility' for the five elements of safety is a good start; at the same time, the number of detail design points are a result of the complexity of the PSE and therefore open the door to different safety levels for different PSEs and the efficient safety management of the total structure.

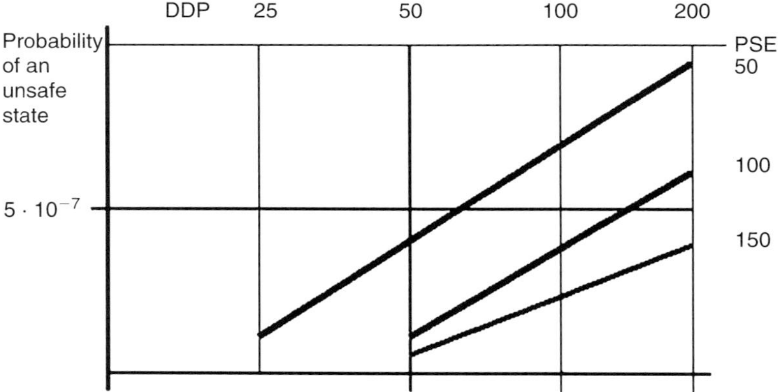

Figure 16.2 Probability of an unsafe state.

16.5 SAFETY STRATEGY AND OBJECTIVES

Returning to the fact that in most cases aircraft structural safety is dependent on both the number of DDPs and the number of PSEs, Figure 16.2 illustrates their effects. It deserves mentioning that ultimate safety, because of the damage tolerance selection of damage size for limit integrity, was assumed to be 15%+, making widespread process failure possible without loss of fail-safe integrity.

It is often the case that loss of ultimate structural properties comes in company with loss of residual strength. This situation makes fail-safe design the fall-back requirement in addition to the ultimate B-value and discrete source requirements.

The whole problem with widespread defects due to process failure is a difficult one, leading to material-specific complications that should be part of the considerations in the selection of material and characterization.

The strategy for precise, practical and safety-based designs resurrects the need for realistic ultimate allowable values and failure criteria. The necessity for fail-safe designs particularly justifies B-values for ultimate designs.

The review of the numerical situation in Figure (16.2) shows that 100 DDPs and 100 PSEs yield an overall structural safety requirement of $4 \cdot 10^{-7}$.

The requirement quoted by ex-President Clinton's Commission of Avation Safety and Security as 10^{-6}, and projected to improve to 10^{-7} during the next two decades, clearly needs process quality improvements during that period.

16.6 DESIGN DATA AND INTEGRITY

Structural integrity is a very useful concept in structural design and is closely related to material allowable values and design data. The ultimate integrity is a very important influence in the 'metal world', together with fatigue and fail-safety, the cornerstones of structural design. In the 'composites world', different kinds of integrities are involved, the most critical often being damage tolerance (residual strength).

The basic drivers in determining design loads in transport-category airplanes are the reversal of internal loads (compression competes with tension, etc.), and must be considered in most structure. Wings have $+2.5g$ to $-1g$ maneuver, and the reversal can be as high as 60%. The balancing tail-load has an up to total reversal, the same applying for gust critical fuselage structure. Lateral load cases (rudder maneuver, lateral gust, etc.) have reversal by definition. The next chapter includes an investigation design load with consistency in definition.

16.7 CONCLUSIONS

The influence of integrity on survival between major inspections is critical.

The total impact on structural safety due to effects from all elements of safety must be considered. Metal safety levels identified by ex-President Clinton's Commission on Safety and Security are such that if panel B-level allowable data and B-level residual strength design data are used in the design, the requirements are totally transferable to composites structural design, if supported by Six-Sigma process quality for the elements of safety.

The improvement in safety level by one order of magnitude during the next decade to keep the total number of fatalities in check, as projected by the same Commission, however, requires fundamental breakthroughs.

A consistent set of design data requirements and failure projections are required to define design data in conventional terms.

The consideration of internal load reversals is the way to produce safe designs.

Fail-safe internal loads must be decreased below ultimate because the structure has a very limited survival time when ultimate stresses are reached.

Chapter 17
Strategy and Objectives for Design

The successful completion of safe composite design, especially in the prevailing environment of innovation, requires a consistent set of design data and a reliable design process. The link between design data and integrity is essential for safety.

17.1   ULTIMATE INTEGRITY

Practice in the 'composites world' has gravitated toward the following deterministic definition:

$$P(\bar{U}_U) = P(\bar{B}_U \mid S_L S_D S_E S_S) \tag{17.1}$$

Here the following events are included:

$\bar{U}_U$: loss of ultimate integrity
$\bar{B}_U$: strength, S is less than ultimate requirement USR
S_L : load producing USR
S_D : concurrent state of damage
S_E : concurrent state of environment (e.g. temperature)
S_S : defect in local load distribution (e.g. open-hole effect).

In order to consider the fact that these events are random the state variables in Equation (17.1) can be rewritten as

$$P(\bar{U}_U) = P(\bar{B}_U \mid S_L S_D S_E S_S) \cdot P(S_L \mid S_D S_E S_S) \cdot P(S_D) \cdot P(S_E) \cdot P(S_S) \tag{17.2}$$

The interpretation of Equation (17.2) leads to the following:

- The first term represents the allowable value definition, e.g. B-value. The given events are: 1.5 · (the limit load we are designing for); existing state of damage at the location; existing state of environment at the location and state of load (e.g. RT or −65F); existing disturbance in local load due to fastener installation defect.
- The second term describes the probability that the limit load is concurrent with the three 'states' identified.

- The third to fifth terms describe the independent probabilities of the three states. The case that is often used includes maximum temperature. We now rewrite Equation (17.2) to consider the case when maximum temperature is not concurrent with an identified load condition:

$$P(\bar{U}_U) = P(\bar{B}_U \mid S_D S_S S_E S_L) \cdot P(S_D) \cdot P(S_S) \cdot P(S_E) \cdot P(S_L) \qquad (17.3)$$

The numerical probabilistic evaluation yields:

$$P(\bar{U}_U) = 10^{-1} \cdot 0.2 \cdot 10^{-2} \cdot 0.1 \cdot 0.1 \cdot 10^{-9} \cdot 0.33 \cdot 10^{-4} = 0.66 \cdot 10^{-19}$$

The probability is very remote, and in comparison with what has been said in previous chapters, it does not have any noticeable effect on structural safety (compare $O(10^{-7})$). One could with reason raise questions about the value there: of as part of design requirements.

It is appropriate to discuss the situation with internal load reversal, and the fact that competing compression- and tension-dominated load conditions makes the criticality comparisons between 'open-' and 'filled-holes' far from a meaningful pursuit.

The design approach to 'internal load reversal' and the same location would have to have process assessments for both 'net fit installations' and 'clearance installation'.

It also seems that the design situation defined above logically would consider the damage that is survivable for ultimate design. The conclusion then will be that, by present practices, the following ultimate design situations must be considered in accordance with Equation (17.3):

- distortions in internal loads due to fastener installation;
- defects in tension, compression and shear;
- compression after impact;
- tension after impact;
- shear after impact;
- buckling after impact;
- crippling after impact.

The random view of design constraints is helpful in selecting a rational combination of allowable values, design states and load cases.

17.2 STATE OF DAMAGE FOR ULTIMATE CONDITIONS

The design for ultimate loads (producing a safety factor of 1.5 for limit loads), and the actual reaching of ultimate internal loads and stresses when fail-safe fall-back is 'activated' makes one wonder why buckling and crippling for panels are not based on B-values with damage included, particularly as the philosophy having evolved for the design of composites glibly states: 'If you cannot see the damage, the structure must be safe with damage included'. Equation (17.3) could be considered the standard for different failure modes, if B_U incorporates criteria for quasi 3-D internal loads states.

We have discussed the problems occurring in relation to widespread toughness, ultimate strength defects and the need for fail-safe design in previous chapters. We suggested the tentative solution of using limit damage sizes so that we get a 15% safety buffer for fail-safe. An alternative for that solution could be a quality control capability that assures detection with a high probability when the tentative solution is awkward.

The need for fail-safe detail designs can have significant safety benefits and, with the right quality control capability for detection of widespread defects, there would not be a price to pay in structural weight.

17.3 FAIL-SAFE INTEGRITY

The regulations require fail-safe capability for the use of B-allowable values for all structures. This requires the demonstration of successful internal load redistributions under limit external load, including compression, which would then be linked to damage resistance.

Because the survival (i.e. successful redistribution of internal loads without total failure) of limit load internal load redistribution (inc. compression), this is a very important compliance test requirement, and a very sensitive process control function must be in place to avoid defects that compromise successful internal load redistribution. The probability of lost fail-safe integrity without detection can be written as

$$
\begin{aligned}
P(\bar{S}_{FS}) &= P\{(\bar{B}_{LT}\,\bar{X}_t D_{5T}\bar{H}_T\bar{Y}_{T1}) \cdot [(R_{DT1}\bar{B}_{LT1})\cup \bar{R}_{DT1}]\}\\
&= P(\bar{R}_{DT1}\mid \bar{Y}_{T1}\bar{B}_{LT}\bar{H}_T\bar{X}_t D_{5T}) \cdot P(\bar{Y}_{T1}\mid \bar{H}_T\bar{B}_{LT}\bar{X}_t D_{5T})\\
&\quad \cdot P(\bar{B}_{LT}\mid \bar{H}_T\bar{X}_t D_{5T}) \cdot P(\bar{H}_T\mid \bar{X}_t D_{5T}) \cdot P(D_{5T}\mid \bar{X}_t) \cdot P(\bar{X}_t)\\
&\quad + P(\bar{B}_{LT1}\mid R_{DT1}\bar{Y}_{T1}\bar{B}_{LT}\bar{H}_T\bar{X}_t D_{5T}) \cdot P(R_{DT1}\mid \bar{Y}_{T1}\bar{B}_{LT}\bar{H}_T\bar{X}_t D_{5T})\\
&\quad \cdot P(\bar{Y}_{T1}\mid \bar{B}_{LT}\bar{H}_T\bar{X}_t D_{5T}) \cdot P(\bar{B}_{LT}\mid \bar{H}_T\bar{X}_t D_{5T})\\
&\quad \cdot P(\bar{H}_T\mid \bar{X}_t D_{5T}) \cdot P(D_{5T}\mid \bar{X}_t) \cdot P(\bar{X}_t) \qquad\qquad (17.4)
\end{aligned}
$$

Here there are two terms: one where limit capability is not restored because the remaining structure does not have internal ultimate capability, and the second where the internal load redistribution failed (damage resistance was not up to requirements). The next example illustrates the numerical values driven by safety requirements.

Example 17.1 Equation (17.4) is used to illustrate the relative values associated with composite fail-safety. Two different scenarios are considered: redistribution failed (damage resistance inadequate); or ultimate strength of 'the remaining structure' does not satisfy the ultimate requirement.

$$P(\bar{S}_{FS}) = 0.10 \cdot 0.10 \cdot 0.10 \cdot 10^{-3} \cdot 10^{-3} \cdot 10^{-2}$$
$$+ \, 0.10 \cdot 0.9 \cdot 0.10 \cdot 0.10 \cdot 10^{-3} \cdot 10^{-3} \cdot 10^{-2}$$
$$= 10^{-11} + 0.9 \cdot 10^{-11}$$
$$P(\bar{S}_{FS}) = 1.9 \cdot 10^{-11}$$

The ultimate 15% positive margin of safety achieved by damage tolerance criticality will make it possible to reduce the second term to zero, again illustrating the 'benefit' of the potential of damage tolerance as a critical mode.

$$P(\bar{S}_{FS}) = 10^{-11}$$

If instead the situation was ultimate strength critical and the damage resistance of the 'remaining' structure would be reduced by 15%, the situation becomes:

$$P(\bar{U}_{FS}V_{23}T_2\bar{X}_M) = P(\bar{U}_{FS} \mid V_{23}T_2\bar{X}_M) \cdot P(V_{23} \mid T_2\bar{X}_M) \cdot P(T_2 \mid \bar{X}_M) \cdot P(\bar{X}_M)$$
$$= 0.82 \cdot 0.2 \cdot 0.5 \cdot 10^{-9} = 0.82 \cdot 10^{-10}$$

and considering the difficulty in detecting this type of defect

$$P(\bar{S}_{FS}) = 0.8 \cdot 10^{-10} \cdot 10^{-1} = 0.8 \cdot 10^{-11}$$

And we can again demonstrate the balance in values between design and other elements in requirements (e.g. manufacturing).

The preservation of fail-safe integrity is very important for composite structure, because of damage tolerance in compression. It makes damage resistance the driver both in damage size and in residual strength. The safety influence makes it necessary to conduct a compliance demonstration program that covers fragments from engine failure, hailstones in flight, 'exploding' tires and bird strikes. The similarity between the nature of fail-safe and discrete source detail design requirements make these two types of integrity much more of a safety threat in composite structure than in metallic. The next section discusses the main aspects of discrete source damage design.

17.4 DISCRETE SOURCE DAMAGE INTEGRITY

Discrete source damage is a very important threat to composite structural safety and involves both damage resistance and damage tolerance. New threats are emerging; for example hailstone impact in flight has become a not infrequently reported phenomenon, and bird strike has continued to be a troublesome event, especially at lower altitudes, even though the required 'get-home' loads provide some relief when experienced pilots are involved.

It appears to be the case that a compliance demonstration requirement should include a focus on maximum damage size. The front pressure bulkhead of the fuselage, especially, but not exclusively, should be made safer by requiring that the damage size allowed be based on the presence of crew and systems in the damage area. This should be complemented with a residual strength requirement at limit load after the damage event.

The common joint event of bird strike can be represented by the following probability expression:

$$
\begin{aligned}
P(\bar{U}_{BS}) &= P(\bar{Y}_{BS} M_{BS} V_{BR} D_{BS} \bar{B}_{BS}) \\
&= P(\bar{B}_{BS} \mid D_{BS} \bar{Y}_{BS} M_{BS} V_{BR}) \cdot P(D_{BS} \mid \bar{Y}_{BS} M_{BS} V_{BR}) \\
&\quad \cdot P(M_{BS} \mid V_{BR} \bar{Y}_{BS}) \cdot P(V_{BR} \mid \bar{Y}_{BS}) \cdot P(\bar{Y}_{BS})
\end{aligned}
\tag{17.5}
$$

The joint event describing the design situation for bird strike can be described as follows, based on Equation (17.5):

The first term of the right hand-side the probability of lost limit load integrity after impact with a stabilized damage size of, D_{BS}, given together with the definition of bird strike.

The second term: the probability of damage size, D_{BS}, given the bird-strike particulars.

The third term: the probability of the mass of the particular bird, given relative speed at impact.

The fourth term: the probability of the relative bird speed, given a strike.

The fifth term: the probability of bird strike.

Example 17.2 The numerical interpretation of Equation (17.5), the probability of loss of bird-strike integrity assuming that design data are available, yields

$$
P(\bar{U}_{BS}) = 10^{-1} \cdot 0.1 \cdot 0.5 \cdot 10^{-1} \cdot 0.4 \cdot 10^{-2} = 2 \cdot 10^{-6}
$$

It is interesting to note that when damage tolerance is made critical, the value for the probability of lost integrity is of the same order of magnitude as for loss of integrity due to bird strike. The remaining question of interest, probability of failure when bird-strike integrity is lost, can be written as

$$P(\overline{A}\,\overline{U}_{BS}) = P(\overline{A}\,|\,\overline{U}_{BS}) \cdot P(\overline{U}_{BS}) = 5 \cdot 10^{-3} \cdot 2 \cdot 10^{-6} = 10^{-10}$$

So assuring that the loads are less than those borne during the rest of the flight is a successful strategy, but requires damage resistance (damage size vs. bird strike) and residual strength vs. damage size, but both must survive the bird-strike event with associated dynamic effects and the get-home loads that should not be greater than the internal loads during the strike event, which can exceed 70% of limit.

17.5 CONCLUSIONS

A logical review of the random nature of the allowable B-value states is needed.

A review is required of the consistency of only considering ultimate damage in B-allowable values, under similar circumstances at the same location.

The importance of fail-safe design for composite structure is clear, particularly the associated problems, and widespread defects due to processing failures make it an imperative even when high-tech quality control processes are employed.

Discrete source damage design compliance and the detail requirements associated with failure under load should be based on point design testing and an approach to capture successful design requirements and data assuring a learning process and service experience of greater efficiency than is the case in the 'metal world'. A key part of this activity is the need to understand damage resistance of composites (new versions) and to create an efficient approach.

Part II
Safety Management Summation

Part I was an exploratory investigation of safety management, elements of safety, safety requirements and compatibility of different safety effects, and alternative views of the nature of the different processes involved, their different roles being contrasted with safety requirements.

Composite structural design and manufacturing and many of the other elements of safety have a distinct random nature both intrinsically and in application. A random world does not easily lend itself to a deterministic interpretation and conceptual difficulties often arise in requirements formulation and interpretation.

Part II will assemble all the pieces investigated in Part I with emphasis on preserving the essence of engineering practices, avoiding overwhelming data requirements and focusing on a practical structural safety approach.

Chapter 18
Elements of Safety and Engineering Practices

The investigation started with a set of elements of safety that have a significant influence on safety issues. The initial set of elements of safety for composites is as follows:

- structural composites design;
- composites manufacturing;
- composites maintenance;
- composites aspects of operation;
- composites formulation of requirements.

There was also a set of secondary elements evolving from the investigation. They are:

- uncertainty management;
- structural monitoring of uncertainty and risk;
- structural monitoring and analysis for updating;
- feedback to education and to service experience database;
  and the
- design process, also added to the total safety management process.

The design process alluded to is the one that focuses on design criteria, design date and allowable values, and test programs in support of structural composites design development. The common effects of defects that have been investigated involve:

- reduction of structural properties;
- increase of internal limit loads;
- spurious mechanical damage.

The attack on safety has been identified as effects of defects that threaten:

- ultimate integrity;
- limit integrity;
- fail-safe integrity;
- discrete source damage integrity;
- damage resistance integrity.

The safety requirements that all of these processes and elements are tested against are discussed in the next section. The measures of safety that are used for design are 'the probability of undetected loss of integrity' and the quality of the processes involved is based on 6σ with nominal effects of defects included. The process is described in Figure 18.1 and the acceptable defect values are shown, while the rest are outside the 6σ range.

The overall results of the exploration involve acting on the fact that all the elements have a clear, random nature and only sometimes can be substituted with concepts such as allowable B-values.

Manufacturing processes of composite parts and assemblies provide one of the more difficult problems: widespread defects affecting both ultimate strength and residual strength. As reduced ultimate properties cause loss of fail-safe integrity and as reduced residual strength requires fail-safe capability the choices are limited. Either quality control must be such that the probability of missing widespread defects is minute or the definition of limit damage sizes must be such that there is an ultimate margin-of-safety buffer of 10–15% that can preserve fail-safe integrity. The ultimate allowable B-values must have a consistent definition of the type.

$$P(\bar{U}_U) = P(\bar{B}_U S_D S_E S_I S_{LU}) = P(\bar{B}_U \mid S_D S_E S_I) \cdot P(S_D) \cdot P(S_I \mid S_E) \cdot P(S_E) \qquad (18.1)$$

where load cases and environments are compatible. The first factor represents the B-value for:

- buckling;
- crippling;

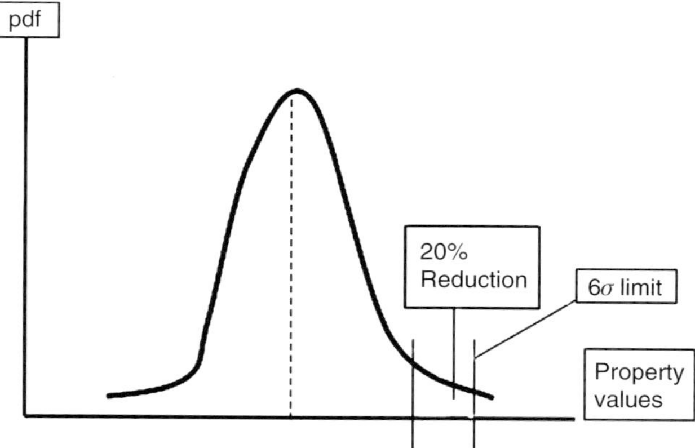

Figure 18.1 Process description.

- compression after impact;
- tension after impact;
- compression with fastener effects;
- tension with fastener effects;
- shear after impact;
- shear with fastener effects.

This list represents the modes of failure, especially with internal load reversal, and with the same damage present for all the 'modes of failure'. The situation with fail-safe as back-up to damage tolerance raises the question of different modes of failure and different damage locations at ultimate internal loads in the remaining structure (after load-path failure), and what the critical combination of failure mode and damage location is. The ultimate full-scale 'destruct-test', at the most, can show one location for each test and the rest only represent 1.5 times limit load, which may not be the critical situation. Thus fail-safe design for composites needs many special considerations from a safety stand-point.

Safety requirements, at specified damage size for limit load requires a probability of non-detection of 10^{-3} and a probability of occurring of 10^{-3} given that damage is present.

With these specifications, B-values for residual strength are required.

$$P(\bar{S}_T) = P(\bar{H}_\tau \bar{U}_T \bar{H}_T) = P(\bar{H}_\tau) \cdot P(\bar{H}_T \mid \bar{U}_T) \cdot P(\bar{B} \mid \bar{X}_T D_5) \cdot P(D_5 \mid \bar{X}_T) \cdot P(\bar{X}_T)$$

The required values are shown below:

$$= 10^{-2} \qquad \cdot 10^{-3} \qquad \cdot 10^{-1} \qquad 10^{-3} \qquad 10^{-2}$$

Interpreted as:

$$P(\bar{H}_\tau) \le 10^{-2}$$
$$P(\bar{H}_T \mid \bar{U}_T) \le 10^{-3}$$
$$P(\bar{B} \mid \bar{X}_T D_5) \le 10^{-1}$$
$$P(D_5 \mid \bar{X}_T) \le 10^{-3}$$
$$P(\bar{X}_T) \le 10^{-2}$$

and satisfying the requirement that $P(\bar{S}_T) \le 10^{-11}$.

18.1 SAFETY BACKGROUND

A review of safety of aviation by ex-President Clinton's Commission on Safety and Security of Aviation was carried out during the second half of the 1990s. It reported a good safety record for transport-category aircraft and recommended maintaining present objectives, one unsafe flight 100000 with an improvement rate of 5% every 20 years to keep the number of casualties from increasing. This was confirmed by the FAA in early 2007.

The validation of safety elements and the quality applied in Part I consequently was:

$$\text{Airplane: probability of one unsafe flight during } 10^5 \text{ flights } 10^{-5}$$

$$\text{Structural: probability of one unsafe flight in } 10^5 \text{ flights } 10^{-6}$$

Structural probability is based on the sum of the probabilities of the number of PSEs and the determination was applied to a PSE as the basic safety unit with calculations conducted at all the DDPs using a model that is based on the nature and geometry of the finite element model used for the internal loads calculations.

The resulting conclusions of Part I were of two kinds. The minimization of the probability of an unsafe state was quite workable in relation to composite structure by identifying detail requirements needed in the design criteria. The solution to widespread undetected defects requires a damage tolerance requirement that produces a buffer in ultimate margins of safety up to the range of 10–15%. The definition of 6σ also requires the inclusion of a range of acceptable defects.

The objective of 6σ quality is to produce a distributed responsibility for safety with even burden. Part I shows a flexible opportunity that satisfies both order of magnitude and a practical use of 6σ processes that add some defect 'sizes' of between 6σ and 7σ.

Therefore the safety requirements are easily satisfied with a reasonable number of PSEs, but do not seem to respond well to safety improvements that are not due to new technologies in materials and processes. Thus the conclusion is that 6σ process quality is a safety requirement that could satisfy the above structural safety requirement if combined with rigorous process control. The nature of both material characteristics and process quality lead to a case-by-case situation and the safety levels are best produced by stringent process control. The sensitivity to defects is a natural variable in material and process choices and development, and a very effective way to establish and maintain safety.

There are a number of situations that are difficult to resolve but which can be given a limited solution. Widespread defects involving ultimate strength and residual strength produce widespread violations of fail-safety which is the only fall-back for fracture mechanics-based damage tolerance.

One limited solution is the buffer of ultimate margin of safety produced when the structure is damage tolerance critical. That ultimate margin makes it possible to accept a 'limited' reduction of ultimate properties.

The margin of safety can be written as:

$$MS = \frac{F_{All}}{f_a} - 1 \Rightarrow F_{All} = (1 + MS)f_a$$

A reduction of the ultimate property by a factor '*k*' yields for example

$$k \cdot F_{All} = k \cdot (1 + MS) \cdot f_a$$

Thus, the reduced allowable value, '$k \cdot F_{All}$', would be equal to the applied stress, 'f_a', if the factor

$$k \cdot (1 + MS)$$

were equal to one.

Example 18.1   We will concentrate on the case $k = 0.8$, which requires

$$(1 + MS) = \frac{1}{0.8} \Rightarrow MS = +0.25$$

We now assume we have an ultimate compression of N_u and an ultimate allowable of

$$F_u = 0.0042 \cdot 12 \cdot 10^{-3} \text{ yielding } t_u = N_u / 50, \text{ and}$$

$$RS = 0.0035 \cdot 12 \cdot 10^{-3} \quad t_L = N_u / 1.5 \cdot 42$$

$$\frac{t_L}{t_u} = \frac{63}{50} \approx 1.25 \Rightarrow MS = +25\%$$

which would allow the ultimate strength to be reduced by a factor of 0.8 and fail-safe integrity maintained. This would mean that all situations involving both ultimate strength and residual strength could be handled in the same way.

The complete answer to the requirement of process safety is:

Six-Sigma process quality with an 'acceptable' defect size (e.g. 10% property reduction) is determined on a case-by-case basis during process development. The only alternative is process control being much more reliable in order to eliminate widespread defects and combined ultimate strength and toughness-reduction, trading weight for cost.

Process development and the selection of materials will determine the relation between defects and property reductions and those reductions that do not occur in combination, but if not explicitly excluded from the situation at one DDP, the design must address the following modes of failure involved in loss of fail-safe integrity (ultimate integrity) in compression:

- compression with damage and fastener effects (open to filled-hole effects);
- buckling with damage and fastener effects;
- crippling with damage and fastener effects.

The general format for design data (allowable values) is

$$P(\bar{U}_U) = P(\bar{B}_U \mid S_D S_E S_I) \cdot P(S_D) \cdot P(S_I \mid S_E) \cdot P(S_E) \tag{18.2}$$

Here several modes of failure are competing, and the state of damage is the same for all load cases; the states of environment are concurrent and compatible. Provision of data to produce fastener installation corrections for MRB action for the appropriate environment is part of what is required for safety. The probability of an unsafe state is

$$P(\bar{U}_U F_K \bar{H}_0) = P(\bar{H}_0 \mid \bar{U}_U \bar{F}_K) \cdot P(\bar{U}_U \mid \bar{F}_K) \cdot P(\bar{F}_K) \tag{18.3}$$

Here undetected loss of integrity due to manufacturing defects is calculated from Equation (18.2), assuming $P(S_I \mid S_E) = 0.1$.

Example 18.2   The contributions per the above list of failure modes are

$$P(\bar{S}_{10}) = 10^{-1} \cdot 0.82 \cdot 0.33 \cdot 0.2 \cdot 10^{-9} = 0.5 \cdot 10^{-11}$$

$$P(\bar{S}_{20}) = 10^{-1} \cdot 0.99 \cdot 0.33 \cdot 0.4 \cdot 10^{-9} = 0.1 \cdot 10^{-11}$$

$$P(\bar{S}_{30}) = 10^{-1} \cdot 0.99 \cdot 0.33 \cdot 0.3 \cdot 10^{-9} = 0.1 \cdot 10^{-11}$$

This numerical illustration again validates the order of magnitude for 6σ process quality. Detection here is based on quality control, in turn based on data for reduced properties with a local definition. The values shown are a result of identifying the most critical locations of damage applying to all modes.

The general list would, in the case of internal load reversal, include

- tension with damage and fastener effects (open to filled-hole effects)

and, in special cases:

- shear with damage and fastener effects (open to filled-hole effects).

This requirement is cumbersome, but lends itself well to the development of an algorithm for producing sizing at every DDP.

It also could be adapted to the concurrent existence of practical environments and design load cases. To the extent maximum temperature, moisture equilibrium conditions and load cases at hand can co-exist, it could be handled by use of the same algorithm, provided a crisp definition is produced for maximum temperature.

18.2 INITIAL INFORMATION AND UPDATING

The initial information required to complete the design comprises probabilities of damage sizes and probability of detection (inspection technology). A monitoring process utilizing service and inspection data combined with updating of probability of damage sizes will assure a continually improving picture of damage probability (Bayesian updating).

An initial set of detection probabilities can be developed from state-of-the-art inspection technology in relation to safety requirements. This set will be in agreement with safety requirements and inspection periods.

The data acquired through the monitoring and analysis process will focus on damage sizes beyond ultimate size requirements. The data developed will cover specific points in time and continually updated records.

The probability of damage being present beyond ultimate size is a good indicator of how well damage accumulation is controlled. These records should compare PSEs from different airplanes of the same model and target PSEs for the same airplane. This will be expanded in order to allow comparisons with different models over time. These results will be used in the Bayesian probability updating, and will also be used in risk management for predicting failure after loss of structural integrity and for scheduling the next inspection accordingly.

The data must contain the size regions D_3, D_4, D_B, D_5 and D_6 and failed load-paths for fail-safe review. Comparison in time is needed for identification of changes in damage accumulation and the presence of hidden damage, to determine the timing for further inspections. Finally, keeping up-to-date data for probabilities of detection for different damage sizes and characteristics serves the purpose of reinforcing the insights in to the probability distribution of detection data of different damage sizes.

This process supports the following data acquisition objectives:

- probability of detection of different damage sizes;
- selection of inspection periods;
- probability of different damage sizes;
- management of risk by major inspection scheduling;
- reduction of uncertainty;

- frequency of load-path failures (inc. unacceptable damage sizes);
- finding defects producing aberrant behavior;
- service databases for education and service experience.

18.3 THE 'WORLD OF COMPOSITES'

Composites are in a state of innovation. New materials, new processes and new structural concepts steadily pass by in review, and service experience and empirical rules are slow to evolve. A transport-category aircraft has yet to be produced with explicit safety constraints, which is an absolute requirement for families of materials and structures with continually varying properties, criticalities and behaviors. FAR and JAR 25 do not contain any comprehensive, composites-specific regulations. The lack of empirical, authoritative databases makes safety-based approaches to composite structural design an absolute must. Now is a good time to make this happen because drift and scatter has become the norm in composite design practices.

The elements of safety have requirements formulation as one of the members there of. The main reason for this is the frequent claim:

Existing regulations are quite applicable to composite structure.

This, however, has turned out not to be true. For example the definition of limit load is not sufficiently stringent, it should be based on load-case and give a random definition of occurrences, and it should describe internal loads, particularly as composites are often critical with secondary loads included.

The present definition 'The largest load expected in service … ,' identifies neither the number of occurrences of limit load during a 'lifetime', nor the probability of reaching limit load, nor what loads are involved.

The role of maximum temperature and its definition of probability are needed for temperature considerations in composites design.

The definition of allowable values must be much more inclusive and related to design load cases as it involves temperature, damage, fastener effects, moisture content and the effects of defects produced by all of the elements of safety.

The list is long and this is just a small portion, but it no doubt makes the case. Composites must have specific regulations for transport-category airplanes. The regulations should identify the structural safety level that has to be met and the consequences, which include allowable B-values, panel buckling and residual strength, should be described.

The general design of composites is based on, at the least, internal loads that are biaxial with shear components. So strength and buckling failure criteria must be part of the requirements, otherwise a design specific test program must be conducted.

18.3.1 *A priori probabilities*

The initial investigation shows that the Six-Sigma process distributions can be a good order of magnitude in terms of process quality for manufacturing defects requirements, if complemented with a gradual transition of 'defect sizes' between 6σ and 7σ. Manufacturing process defects can cause a reduction of structural properties. Widespread defects were shown to be a serious threat to fail-safe design, and are best handled by avoidance. The development of the required processes will provide information on how to group properties.

We often find that compression, shear and tension strength behave differently, but the behaviors of compression strength, stiffness and toughness are similar. The same is true for shear, and tension is the 'odd man out'. It is very important during process development and characterization to resolve these sensitivities, which otherwise can result in separate detail for each structural property. The following format is the governing principle:

$$P(\bar{U}_U \bar{F}_M) = P(\bar{U}_U \mid \bar{F}_M) \cdot P(\bar{F}_M)$$

where the last factor can be written for manufacturing as

$$P(\bar{F}_M) = P(V_{ij} T_j \bar{X}_M) = P(V_{ij} \mid T_j \bar{X}_M) \cdot P(T_j \mid \bar{X}_M) \cdot P(\bar{X}_M) \tag{18.4}$$

Definition:

V_{ij} : extent of defect (e.g. up to 15% reduction)
T_j : type of defect (e.g. compression strength reduction)
$\bar{X}_M$: defect is present.

A numerical illustration yields

$$P(\bar{F}_M) = 0.3 \cdot 0.2 \cdot 10^{-9} = 6 \cdot 10^{-11}$$

The characterization of the process will yield the first two factors. A general description of a DDP yields the following list of participating modes of failure, especially in terms of internal load reversal and proximity of structural interfaces:

- compression with damage and fastener effects;
- shear with damage and fastener effects;
- tension with damage and fastener effects;
- buckling with damage and fastener effects;
- crippling with damage and fastener effects;
  and, when damage tolerance critical,
- reduction in toughness.

Residual strength reduction in relation to limit damage sizes involves damage resistance, damage growth rates and damage tolerance, and might of necessity involve fail-safe fall-back positions.

The other significant influence deriving from defects comes from exceedances of internal limit load due to defects in geometry at installations of fasteners, variations of bond-lines and grip-lengths or faulty torque.

These types of defect are best guarded against to avoid failure, through taking up the fail-safe fall-back position, for stepwise load increases of 10% and upward, which otherwise can cause local failures in less than a realistic inspection interval (<1000 flights).

Many of the same facts apply to defects due for example to improper maintenance causing load increases in design loads. Thoughtful safety management would involve inspection both of the maintenance approach and of the state of the structure.

Finally, the rules for both maintenance and operation should mandate yearly training of key personnel because defects in maintenance and operation can cause difficulty to anticipate many of those defects and counter in the design.

18.3.2 Composites safety management

The probability of an unsafe structure pertinent to safety management has already been modified to include the processes that produce regulations, requirements, design criteria and testing procedures, because they have huge influence on safety levels, especially when they evolve from local practices.

Accidental damage is almost synonymous to composites, so the definition of the probability of an unsafe state is upgraded with the event accidental damage. Analogously to the approach we followed in the earlier development of the expression for the probability of an unsafe structure we arrive at:

$$P(\bar{S}) = P(\bar{D} \mid MIOR\bar{A}_D) + P(\bar{M} \mid IOR\bar{A}_D) + P(\bar{I} \mid OR\bar{A}_D) \\ + P(\bar{O} \mid R\bar{A}_D) + P(A_D \mid R) + P(\bar{R}) \tag{18.5}$$

Here the following events apply:

$\bar{S}$: unsafe state
$\vec{D}$: unsafe design
M : safe manufacturing
I : safe maintenance
O : safe operation
R : safe requirements
$\bar{A}_D$: no accidental damage.

We can identify the following terms in Equation (18.5):

The first term represents the probability of unsafe design, given safe manufacturing, safe maintenance, safe operation, safe requirements and no accidental damage.

The second term is the probability of safe manufacturing, given safe maintenance, safe operation, safe requirements and no accidental damage.

The third term is the probability of safe maintenance, given safe operation, safe requirements and no accidental damage.

The fourth term is the probability of safe operation, given safe requirements and no accidental damage.

The fifth term is the probability of accidental damage, given safe requirements.

The sixth term is the probability of unsafe requirements.

It is interesting to note that this formulation leaves room for dealing with discrete source damage such as engine burst, bird strike, ground debris, hailstones in flight, tire burst and perhaps ground collisions – mainly 'acts of God' – while mishaps suffered by airplanes categorized under a particular element of safety are recorded taking into account the particular element in question.

18.4 CONCLUSIONS

Empirical methods in composites design are not available, in particular for transport-category airplanes.

Regulations that apply to metals rarely do so to composites.

Requirements formulation is included among the elements of safety because of their importance to safety.

Part III
Improved Structural Safety of Composites

INCREASED STRUCTURAL SAFETY FOR THE FUTURE

The analysis and recommendations by ex-President Clinton's Commission on Safety and Security in Aviation established that the present safety levels of transport-category aircraft in the 'metal world' are excellent as they are, and should initially be replicated in the 'composites world'. However, the need to be improved by at least one order of magnitude during the next few decades was supported. The FAA confirmed their support of this projection, because it will keep the number of fatalities constant, while the rates would decrease. The objective is stated thus: 'The number of airplanes is expected to increase, so the rates have to go down to keep the number of fatalities constant'.

Chapter 19
Improvement in Structural Safety of Composites

Part II showed that structural safety levels meet challenges from all elements of safety and that the natural path to improvement lies with the reduction of the probability of an unsafe state. Examination of the following expression (Equation (19.1)) is a good starting point:

$$P(\overline{S}_L) = P(\overline{A} \mid \overline{S}_L WR) \cdot P(\overline{S}_L \mid WR) \cdot P(W \mid R) \cdot P(R) \tag{19.1}$$

$\overline{S}_L$: an unsafe level of safety
$\overline{S}$: an unsafe state
W : weight of aluminum is greater than weight of composite.

The resulting equation for the case where a given unsafe state makes the first factor tend to a value of 1 when the number of flights increases is

$$P(\overline{S}_L) = 1 \cdot P(W \mid \overline{S}_L R) \cdot P(\overline{S}_L \mid R) \cdot P(R) \tag{19.2}$$

This equation is based on detection before failure (first factor).

The second factor describes the probability that there is weight-saving given an unsafe state and safe requirements (including safe regulations).

The third factor describes the probability of an unsafe state, given safe requirements.

The fourth factor represents the probability that the requirements are safe.

Example 19.1 This example illustrates the situation of 'one order of magnitude improvement'. Equation (19.2) can be evaluated as follows:

$$P(\overline{S}_L) = 1 \cdot 0.8 \cdot 10^{-12} \cdot 1 = 0.8 \cdot 10^{-12}$$

Here the value is based on N PSEs and n DDPs. The safety value for the total structure is 10^{-7}. The value required at a DDP is $10^{-7}/N \cdot n$ for the total number of threats. For one threat we have $10^{-8}/N \cdot n$.

The order of magnitude is illustrated at a DDP in Figure 19.1.

The primary target for improving the level of safety one order of magnitude is to enhance quality control so that widespread defects are detected with a very high probability.

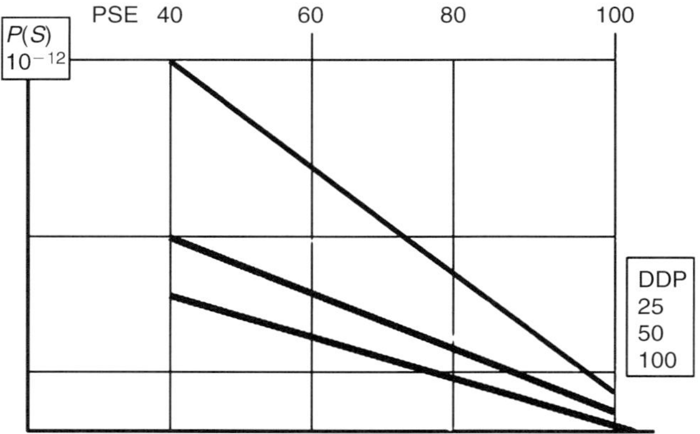

Figure 19.1 Improved safety at DDP.

The second target, the probability of an unsafe state, $P(\bar{S})$

$$P(\bar{S}) = P(\bar{H}_\tau)P(\bar{B}_L \mid \bar{X}D_5) \cdot P(\bar{H}_T \mid \bar{X}D_5)P(D_5 \mid \bar{X}) \cdot P(\bar{X}) \le 10^{-12} \qquad (19.3)$$

as we see must be decreased by one order of magnitude for all processes except one.

The third target is improving the discrete source integrity defects sensitivity, and the design development testing, so that the probability of an unsafe state is improved by one order of magnitude. Finally, the requirements (regulations) for compliance demonstration and the design criteria must be elucidated, because they are addressing a quite a different set of requirements very different from those pertaining to the 'metal world'.

19.1 WIDESPREAD DEFECTS

The solution to the safety problem associated with the present level of safety, i.e. a process that can use the ultimate margin of safety for damage tolerance critical structure in order to compensate, will be an awkward approach for any increases beyond the present level of safety. Thus the requirements for process quality, including detection, would have to be improved. This is covered by the following equation:

$$P(\bar{U}_{FS}\bar{H}_0) = P(V_{ij}T_i\bar{X}_M\bar{H}_0) = P(\bar{H}_0 \mid V_{ij}T_i\bar{X}_M) \cdot P(V_{ij} \mid T_i\bar{X}_M)$$
$$\cdot P(T_i \mid \bar{X}_M) \cdot P(\bar{X}_M) \qquad (19.4)$$

A numerical evaluation of what the projected safety level requires can be found in Example 19.2.

Example 19.2 This example demonstrates the answer to undetected loss of fail-safe integrity according to Equation (19.4):

$$P(\bar{U}_{FS}\bar{H}_0) = 10^{-2} \cdot 10^{-1} \cdot 10^{-1} \cdot 10^{-9} = 10^{-13}$$

Hence the fact that a moderate quality standard of quality control could 'eliminate' the threat of widespread defects is a line of development that is worth pursuing.

19.2 DISCRETE SOURCE DAMAGE INTEGRITY

Discrete source damage in the 'metallic world' has been evolving empirically, and in the 'composites world' a Design Development approach (Point design testing), has been evolving slowly. The guiding discrete damage comes in four categories for transport aircraft. They are impacts relating to:

- bird strike;
- tire (landing gear in stowed position) burst;
- hailstone impact in flight;
- engine burst blade 'penetration'.

To produce a structurally feasible design, one is commonly forced to conduct design development testing and acquire the statistics describing the probability of penetrations that cause collateral damage to crew and systems and hidden damage that threatens the damage tolerance limit integrity. Each of the impacts listed above has its own set of variables, both for data and for design. The unique nature of the items in the list is discussed in the following sections.

19.2.1 Bird strike

A number of situations for a specific thickness in the strike zone pertain to the particular PSE under consideration, but we focus here on specific bird size and relative velocity. First we investigate penetration and failure:

$$P(\bar{A}_P\bar{A}\bar{X}_BMV) = P(\bar{A} \mid \bar{A}_P\bar{X}_BMV) \cdot P(\bar{A}_P \mid \bar{X}_BMV)$$
$$\cdot P(\bar{X}_B \mid MV) \cdot P(V \mid M) \cdot P(M) \tag{19.5}$$

$\bar{A}_P$: penetration of PSE strike zone
$\bar{A}$: failure of PSE
M : bird size (lb)
V : relative velocity of impact
$\bar{X}_B$: bird strike.

$$P(\bar{A}_P \bar{X}_B MV \bar{U}_{BS}) = P(\bar{U}_{BS} \mid \bar{A}_P \bar{X}_B MV) \cdot P(\bar{A}_P \mid \bar{X}_B MV)$$
$$\cdot P(\bar{X}_B \mid MV) \cdot P(V \mid M) \cdot P(M) \tag{19.6}$$

$\bar{U}_{BS}$: loss of bird strike integrity following penetration.

$$P(\bar{X}\bar{U}_{BS}\bar{X}_B MV\bar{H}) = P(\bar{U}_{BS} \mid \bar{X}_B \bar{X} MV\bar{H}) \cdot P(\bar{H} \mid \bar{X}_B \bar{X} MV) \cdot P(\bar{X} \mid \bar{X}_B MV)$$
$$\cdot P(\bar{X}_B \mid MV) \cdot P(V \mid M) \cdot P(M) \tag{19.7}$$

$\bar{X}$: internal damage present
H : damage not detected at post-flight inspection.

We will investigate Equation (19.5) first:

> The first factor on the right hand-side is probability of failure of PSE after penetration.
> The second factor is the probability of penetration with bird size and velocity given.
> The third factor is the probability of velocity given bird size M.
> The fourth factor is the probability of bird size M.

Equation (19.5) describes failure of PSE preceded by PSE penetration. A natural way to improve safety is to reduce the product of the first two factors on the right-hand side by the appropriate number, which requires either a 'better material' or better lay-up.

Equation (19.6) describes loss of bird strike integrity following PSE penetration. A desired improvement can be achieved by decreasing the product of the first two factors on the right-hand side, for example by using a new material a softer lay-up or a hybrid solution.

Equation (19.7) represents detected internal PSE damage. An improvement in safety can be achieved by improving the probability of detection and/or improving the damage resistance of the lay-up.

The first two factors on the right-hand side of Equation (19.5) are produced during design development testing for a specific thickness and lay-up. The following two are produced from the statistics for transport aircraft (which would be a lot more useful if bird size and speed together with damage extent were included).

The first two factors on the right-hand side of Equation (19.6) are produced by design development testing. The last two are acquired from the transport-category statistics.

The first three factors on the right-hand side of Equation (19.7) are produced by design development testing. The last two are part of the transport-category statistics.

The alternative to this approach is an extensive technology program preceding the design stage. In general this is very important for safety and very difficult to resolve without extensive testing.

19.2.2 Tire burst

This is also a difficult problem to solve for composite structure; it is a consequence of aborted take-offs. The landing gears are in stowed position. The heat generated by the brakes causes the tires to burst, and the incident involves the production of a pressure jet and the creation of tire fragments.

Clearly, the severity of the event and the resulting damage depend on its proximity to 'primary structure', e.g. the rear spar, the pressure deck, the primary structure floor beams and the keel-beam, so any design development testing must take into account passenger safety in the cabin following penetration and potentially direct hits from flying debris, and, of course, structural integrity both during and after the event.

The governing equations are as follows:

For penetration and failure: 1.

$$P(\bar{A}_P \bar{A} D T_B) = P(\bar{A} \mid \bar{A}_P D T_B) \cdot P(\bar{A}_P \mid D T_B) \cdot P(D \mid T_B) \cdot P(T_B) \tag{19.8}$$

$\bar{A}_p$: penetration of PSE
$\bar{A}$: failure of PSE
D : distance from tire burst
T_B : tire burst.

> The first factor on the right-hand side of Equation (19.8) is the probability of failure, given tire burst, penetration and distance D away.
> The second factor on the right-hand side is the probability of penetration given tire burst at distance D away.
> The third factor on the right-hand side is the probability of a distance D away given tire burst.
> The fourth factor on the right-hand side is the probability of a tire burst.

For penetration: 2.

$$P(\bar{A}_P T_B D) = P(\bar{A}_P \mid T_B D) \cdot P(D \mid T_B) \cdot P(T_B) \tag{19.9}$$

> The first factor of the right-hand side of Equation (19.9) represents the probability of failure (penetration), given tire burst at a distance of D.
> The second factor of the right-hand side is the probability of a distance D away, given tire burst.
> The third factor is the probability of tire burst.

For loss of tire burst integrity: 3.

$$P(\bar{U}_{TB} T_B D \bar{H}) = P(\bar{U}_{TB} \mid T_B D \bar{H}) \cdot P(\bar{H} \mid T_B D) \cdot P(D \mid T_B) \cdot P(T_B) \tag{19.10}$$

The first factor on the right-hand side of Equation (19.10) represents the probability
of loss of integrity, given tire burst at a distance D away and not detected.

The second term on the right-hand side is the probability of not detecting the damage
from tire burst at a distance D away.

The third term on the right-hand side is the probability that the distance is D given a
tire burst.

The fourth term on the right-hand side is the probability of a tire burst.

The improvement of safety defined for the future will be achieved if the probability of
failure is decreased, if the probability of penetration is decreased or if structure and system protection are added as this involves a confined space.

19.2.3   Hailstone impact in flight

In the last few years there has been an increase in the number of hailstone-related incidents reported. It is tempting to recommend that weather reporting should be improved
so that avoidance would become the main countermeasure. However, composite structure
has the potential for being protected from impact damage in well-defined areas by the
local use of damage-resistant lay-ups. Thus particularly exposed areas could be improved
by using design development testing to achieve better damage resistance.

The governing equation is

$$P(\bar{U}_{DT}) = P(\bar{X}_{ST} M_S V D_{HS} \bar{B}_{HS}) = P(\bar{B}_{HS} \mid \bar{X}_{ST} M_S V D_{HS}) \cdot P(D_{HS} \mid \bar{X}_{ST} M_S V)$$
$$\cdot P(V \mid \bar{X}_{ST} M_S) \cdot P(M_S \mid \bar{X}_{ST}) \cdot P(\bar{X}_{ST}) \quad (19.11)$$

$\bar{X}_{ST}$: hailstone in flight
M_S : hailstone size
V : hailstone relative impact speed
D_{HS} : damage size due to hailstone impact in flight
$\bar{B}_{HS}$: loss of hailstone integrity (strength $S \le$ get-home load).

The first two factors on the right-hand side of Equation (19.11) relate to design development testing phase and an improvement of safety would be produced by the provision
of more damage-resistant lay-ups in the critical zones.

The required statistics would supply values for the last three factors in Equation (19.11).

The expectation is that the FAA will provide the appropriate statistics for transport-
category aircraft of composite structure.

19.2.4   Engine turbine blade penetration

Safety-associated blade penetration is based on fail-safe design. It is critical in identifi-
able zones, where design development testing can be used to demonstrate compliance
with the required 'get-home-load integrity'.

It seems that the safety improvements projected for the future will most easily be realized by making the design load 10% higher than the presently required level.

This discrete case is closely tied to 'pressurized compartment loads' requirements which apply to holes of up to $20\,\mathrm{fl}^2$ between compartments and to the exterior with a relief valve setting of $1.33 \times$ one atmosphere (1.67 above 45000 feet), and 1 g level loads.

It appears that a combined design development program should be considered for the particular type of design development that is required to demonstrate structural integrity. If the required pressurized compartment load can be shown to involve very improbable events, conventional fail-safe design could be used for the demonstration of discrete case compliance with the load increase mentioned above.

19.3 UNSAFE STATES AND SAFETY IMPROVEMENTS

The probability of an unsafe state is a governing quantity for safety design constraints.

The probability of an unsafe state is defined by the following expression

$$P(\bar{S}_T) = P(\bar{U}_T \bar{H}_T) = P(\bar{H}_\tau \bar{B}_I \bar{X}_t D_{5T} \bar{H}_T) = P(\bar{H}_T \mid \bar{X}_t D_{5T} \bar{B}_I) \cdot P(\bar{B}_I \mid \bar{X}_t D_{5T})$$
$$\cdot P(D_{5T} \mid \bar{X}_t) \cdot P(\bar{X}_t) \cdot P(\bar{H}_\tau) \qquad (19.12)$$

Here the factors are:

> The first: the probability of detection at time T
> The second: the probability that residual strength is less than the allowable value
> The third: the probability that the damage size is in D_5
> The fourth: the probability that damage is present
> The fifth: the probability that damage is not detected at the beginning of the time interval.

Example 19.3 Equation (19.12) is now used for a numerical illustration of what it takes to improve from

$$P(\bar{S}_T) = 10^{-11} \quad \text{to} \quad P(\bar{S}_T) = 10^{-12}$$

$$P(\bar{S}_T) = 10^{-11} = 10^{-3} \cdot 10^{-1} \cdot 10^{-3} \cdot 10^{-2} \cdot 10^{-2} \Rightarrow 10^{-12}$$

If the product of the first factor and the third factor becomes 10^{-7}, and as this product represents the 'detection' aspect of undetected loss of damage tolerance integrity, it is a very difficult requirement.

So, Safety improvements in design are therefore seen to be achieved by better damage detection and the more reliable distribution of large damage sizes.

19.4 CONCLUSIONS

The most challenging fields for safety improvements in transport-category airplanes are as follows.

Discrete source damage design to higher safety levels is required, because new data-bases must be established.

Widespread defects protection requires improvements in quality control reliability.

The baseline requires B-value design data for panel and residual strength design data.

Chapter 20

Process Quality and Integrity of Requirements Formulation

The initial choice of process quality, Six-Sigma, is now investigated to determine its adaptability. The objective of the process is to protect structural integrity; the governing equation is

$$P(\bar{U}_I \bar{F}_K \bar{H}_0) = P(\bar{U}_I \mid \bar{F}_K \bar{H}_0) \cdot P(\bar{H}_0 \mid \bar{F}_K) \cdot P(\bar{F}_K)$$
$$= P(\bar{U}_I \mid \bar{F}_K \bar{H}_0) \cdot P(\bar{H}_0 \mid \bar{F}_K) \cdot P(V_{KL} T_K \bar{X}_R) \qquad (20.1)$$

Requirements, criteria and regulations have evolved from developed practices for composites. The subject has turned out to be very important for safety. Regulations – or the lack of them – have set minimum standards, which are therefore the necessary foundation for all safety requirements in composite structures, and should be based on a sound technical foundation for process quality and issued on the basis of justifications and arguments that will be reviewed by the practicing community and updated – based on solid arguments, not a majority vote. This is a high-value technical process, not a bureaucratic one.

The formulation and updating of regulation can be designed as Six-Sigma processes. The process in question is governed by

$$P(\bar{S}_T) = P(\bar{U}_L V_{KL} T_K \bar{X}_R \bar{H}_0) = P(\bar{U}_L \mid V_{KL} T_K \bar{X}_R) \cdot P(\bar{H}_0 \mid V_{KL} T_K \bar{X}_R)$$
$$\cdot P(V_{KL} \mid T_K \bar{X}_R) \cdot P(T_K \mid \bar{X}_R) \cdot P(\bar{X}_R) \qquad (20.2)$$

The sub-events are:

$\bar{U}_L$: loss of integrity, L
V_{KL} : extent of defect
T_K : type of defect
$\bar{X}_R$: mishap in regulation formulation
$\bar{H}_0$: not detected before submission.

This process will be executed for regulation formulation.

20.1 PROCESS FOR REGULATIONS

The 6σ process will now be used for regulation formulation. Equation (20.1) will be modified to fit the situation:

$$P(\bar{U}_I \bar{F}_K) = P(\bar{U}_L \mid V_{KL} T_K \bar{X}_R) \cdot P(V_{KL} \mid T_K \bar{X}_R) \cdot P(T_K \mid \bar{X}_R) \cdot P(\bar{X}_R) \qquad (20.3)$$

The next example illustrates how the use of mean value of strength for allowable design value influences safety.

Example 20.1 Equation (20.1) is an expression for loss of integrity due to a defect in the requirements. The use of mean for design can be expressed as

$$P(\bar{U}_L \bar{F}_K) = 0.5 \cdot 0.5 \cdot 0.5 \cdot 10^{-9} = 1.25 \cdot 10^{-10}$$

Here the first factor of the right-hand side of Eqn. (20.3) is the probability of loss of integrity

The second factor represents the use of the value

The third factor is the probability of this type of 'defect'.

The value of the probability of an unsafe state as calculated above is too large. A comparison with B-values yields

$$P(\bar{U}_L \bar{F}_K) = 0.1 \cdot 0.05 \cdot 1 \cdot 10^{-9} = 0.5 \cdot 10^{-11}$$

Here 0.1 is the B-value characteristic. The second figure represents the acceptable confidence defect for allowable values. The third figure is the probability of this type of situation for B-values. The value $0.5 \cdot 10^{-11}$ is well within an acceptable range of safety.

20.1.1 Sizing requirements and criticality

The sizing of composite structure is much more complicated than the equivalent process in the 'metal world'. Practice has evolved toward an allowables definition, in principle resulting in the approach discussed below. Ultimate allowable values (B-values with fail-safe design) have the following background of integrity:

$$P(\bar{U}_U) = P(\bar{B}_U S_D S_E S_S S_L) = P(\bar{B}_U \mid S_D S_E S_S S_L) \cdot P(S_D)$$
$$\cdot P(S_E \mid S_L) \cdot P(S_S \mid S_L) \cdot P(S_L) \qquad (20.4)$$

$\bar{U}_U$: loss of ultimate integrity
$\bar{B}_U$: ultimate strength is less than ultimate requirement

S_D : state of damage
S_E : state of environment (temperature, moisture, etc.)
S_S : state of local stress field distortion
S_L : set of load cases compatible with the selected states.

The sizing at a detail design point (DDP) is a competition between a number of modes of failure (kinds of allowable values) and a compatible set of states. The competing values are:

- compression with fastener effects and damage;
- buckling with fasteners and damage;
- crippling with fasteners and damage;
- tension with fastener effects and damage (if internal loads reversal);
- shear with fastener effects and damage (if a shear region).

Common states include ultimate damage sizes, temperature (RT, $-65F$ or max. temperature), equilibrium moisture content, and pertinent fastener installation.
    A critical situation is the result of processing sets of different states.
    Equation (20.4) defines the allowable values per

$$P(\bar{B}_{Ui} \mid S_{Di}S_{Ei}S_{Si}S_{Li})$$

Here $\bar{B}_{Ui} \Leftrightarrow \{s \leq USR_i)$, which applies to a specific group of states.
    This approach clearly requires both a criterion for failure and one for buckling interaction with biaxial plus shear internal loads. The defects arising from inadequate design data could easily cause a 10% over-prediction resulting in (for $C_V = 0.1$):

$$\Phi\left(\frac{USR - 1.035USR}{\sigma}\right) = \Phi(-0.33) = 0.37$$

And the result yields

$$P(\bar{U}_U \bar{F}_R) = P(\bar{B}_U \mid V_{23}T_2\bar{X}_R) \cdot P(V_{23} \mid T_2\bar{X}_R) \cdot P(T_2 \mid \bar{X}_R) \cdot P(\bar{X}_R)$$

The numerical illustration for this defect is

$$P(\bar{U}_U \bar{F}_R) = 0.37 \cdot 0.3 \cdot 0.7 \cdot 10^{-9} = 0.8 \cdot 10^{-10}$$

This is an unacceptable value even though the defect is modest.

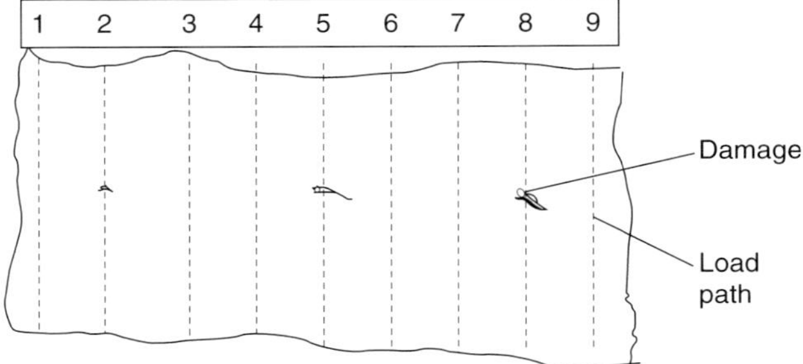

Figure 20.1 Load paths and fail-safe.

20.1.2 Fail-safe design and loss of fail-safe integrity

Figure 20.1 illustrates how the damage indicated causes loss of fail-safe integrity for 1, 3, 4, 6, 7 and 9.

We assume that the probability of loss of fail-safe integrity is

$$P(\bar{U}_{FS}) = P(X_1\bar{X}_2X_3X_4\bar{X}_5X_6X_7\bar{X}_8X_9) = [P(\bar{X})]^3 \cdot [P(X)]^6$$
$$= (10^{-2} \cdot 10^{-1})^3 \cdot (0.999)^6 = 0.99 \cdot 10^{-9}$$

Here six out of nine load paths have lost fail-safe integrity, even with an ultimate margin of safety in place. Is this in actuality fail-safe structure?

It could become so when there is an ultimate margin of safety due to damage tolerance criticality. The next example illustrates the situation.

Example 20.2 A situation of ultimate margin of safety due to one example of a damage size definition is investigated.

Figure 20.2 describes damage size regions. The ultimate region ends at MUD (maximum ultimate damage) and the limit region starts at EDD (easily detectable damage). (NDD is not detectable damage; GDD is good detectability damage; MAD is maximal detectable damage.)

We assume a square-root of damage size ratio relation for residual strength for different damage sizes:

The limit allowable value is $\sqrt{0.33} \cdot USR$

The limit gauge is $t_L = \dfrac{N}{1.5\sqrt{0.33}\,USR} = 1.16\,\dfrac{N}{USR} = 1.16 \cdot t_U$

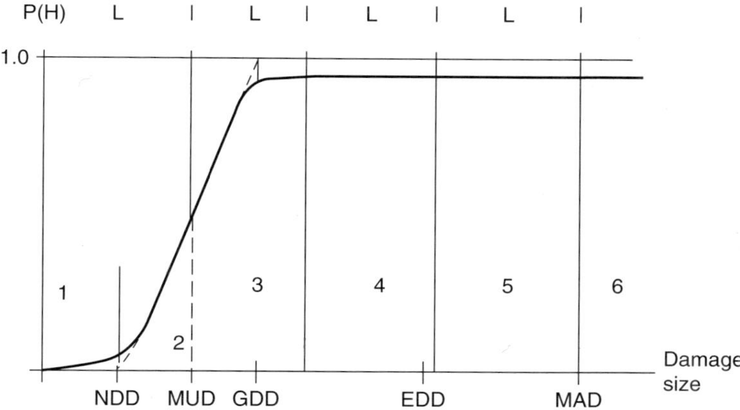

Figure 20.2 Damage size regions.

The consequence is that we have a 16% ultimate margin of safety. The resulting fail-safe integrity would last up to the damage tolerance region. This can be verified when allowable value testing is completed.

20.2 CONCLUSIONS

Correcting the growth of defects between 6σ and 7σ to a gradual growth from 0 to 10% results in an improved value of $0.5 \cdot 10^{-12}$ and a more manageable value for the effects of defects due to the elements of safety.

The rationale will be discussed in Chapter 24; a more realistic value can be achieved for the probability of a total unsafe state, and for the long-term objective of zero defects below 6σ.

Chapter 21
Safety Management and Process Quality

Safety requirements apply to airplane safety, to structural safety and to the safety relating to each individual principal structural element (PSE). Simplicity would lead one to assume an equal requirement for every PSE. So we will assume there are 100 PSEs, which, for example, would lead to a set of safety requirements in terms of probabilities of one unsafe flight per 100 000 flights, yielding:

$$\text{for the airplane: } 10^{-5}$$
$$\text{for the structure: } 10^{-6}$$
$$\text{per PSE: } 10^{-8}$$

For a situation where PSEs can be shown to have equal importance, because of the similar consequences of failure of any PSE, that would be the expected choice. If, however, it were possible to classify the PSEs in three categories as an example, we would have:

1. very serious consequence: 0.90C; 20
2. serious consequence: 1.00C; 30
3. not-so-serious consequence: 1.10C; 50.

The total structure would have the requirement:

$$10^{-6} = (18 + 30 + 55)C \Rightarrow C = 0.0097 \cdot 10^{-6}$$

This would result in:

$$20 \text{ PSEs with the requirement } 0.0087 \cdot 10^{-6}$$
$$30 \text{ PSEs with the requirement } 0.0097 \cdot 10^{-6}$$
$$50 \text{ PSEs with the requirement } 0.0107 \cdot 10^{-6}$$
$$\Sigma \ldots\ldots\ldots\ldots\ldots\ldots\ldots 10^{-6}$$

which would lead to three different requirements for PSEs. If the 'very serious' had 35 detail design points (DDPs) then category 1 would have the requirement $0.025 \cdot 10^{-8}$; if the 'serious' had 30 then category 2 would have the requirement $0.032 \cdot 10^{-8}$; and if the 'not-so-serious' had 25 then category 3 would have $0.042 \cdot 10^{-8}$.

It is evident that one can achieve many levels of safety for the detail design points (DDPs), and in this example we have:

Category 1: $1.25 \cdot 10^{-11}$
Category 2: $1.07 \cdot 10^{-11}$
Category 3: $0.86 \cdot 10^{-11}$

Which are the detail requirements per process per DDP.

21.1 MANUFACTURING PROCESS SAFETY MANAGEMENT REQUIREMENTS

The investigations of the potential for 6σ to be a practical quality for manufacturing processes have been focused on defects that have structural significance, and consequently involve the following effects:

- reduction of structural properties;
- increase of internal structural loads;
- spurious mechanical damage.

21.1.1 *Defects and structural properties*
A typical threat against structural integrity can be written as follows:

$$\Delta P(\bar{S}_i) = P(\bar{U}_I V_{KL} T_K \bar{X}_M)$$
$$= P(\bar{U}_I \,|\, V_{KL} T_K \bar{X}_M) \cdot P(V_{KL} \,|\, T_K \bar{X}_M) \cdot P(T_K \,|\, \bar{X}_M) \cdot P(\bar{X}_M) \qquad (21.1)$$

We now assume that this equation adresses a typical integrity, e.g. ultimate integrity, and that we have a 6σ process containing defects that have an up to 20% reducing effect that acts on B-values.
This results in

$$\Phi\left(\frac{USR - 0.92USR}{0.92 \cdot USR \cdot 0.1}\right) = \Phi\left(\frac{1.087 - 1}{0.1}\right) = 1 - \Phi(-0.87) = 0.81$$

The following example deals with ultimate axial strength.

Example 21.1 Equation (21.1) yields the following contribution to the probability of an unsafe state:

$$\Delta P(S_i) = 0.81 \cdot 0.1 \cdot 0.1 \cdot 10^{-9} = 0.81 \cdot 10^{-11}$$

which yields an order of magnitude in the range shown in the requirement investigation on the previous pages. To achieve lower requirements we could use a 'tighter' process that would include only up to 10% reduction, yielding

$$\Phi\left(\frac{USR - 1.035USR}{1.035 \cdot USR \cdot 0.1}\right) = \Phi\left(\frac{0.97 - 1}{0.1}\right) = \Phi(-0.34) = 0.36$$

Otherwise a more stringent process quality must be adopted.

If we now look at all the structural properties involved, in Table 21.1 we have as an example (validated by testing during process characterization):

The probability range for this example is

$$P(\overline{F}_{Mi}) = P(V_{KL}T_K\overline{X}_M) = P(V_{KL} \mid T_K\overline{X}_M) \cdot P(T_K \mid \overline{X}_M) \cdot P(\overline{X}_M)$$

which produces the largest value: $0.10 \cdot 0.10 \cdot 10^{-9} = 10^{-11}$, and the smallest value: $0.03 \cdot 0.03 \cdot 10^{-9} = 0.9 \cdot 10^{-12}$. The main reason for this format is the concern about coupling, for example the often-discussed case of one defect producing reduction in strength, stiffness and toughness, which is a material or process characteristic.

The total contribution is the sum of all the individual contributions of each of the processes involved. Moreover, Table 21.1 represents an example of a 'contribution from

Table 21.1

Type	$P(V_{KL} \mid T_K\overline{X}_M)$	$P(T_K \mid \overline{X}_M)$
COMPOSITES		
Compression strength	0.10	0.10
Tension strength	0.05	0.05
Shear strength	0.10	0.10
Compression stiffness	0.10	0.10
Tension stiffness	0.03	0.03
Shear stiffness	0.10	0.10
Compression toughness	0.10	0.10
Shear toughness	0.10	0.10
Tension toughness	0.03	0.03
Damage resistance	0.04	0.04
Damage growth rate	0.10	0.10
ADHESIVE		
Shear strength	0.10	0.10
Tension strength	0.10	0.10
Shear toughness	0.10	0.10
Damage growth rate	0.10	0.10

manufacturing'. The case is itself a challenge and requires a strategy in eliminating well-identified defects.

21.1.2 Defects in manufacturing causing increased internal loads

There are two types of increases: errors in geometry that cause self-equilibration and errors in geometry that cause internal load concentration (e.g. mismatched holes). The probabilities can be formulated similarly to what was done for 'reduced structural properties' in the previous section, but this represents a different process. The limit of the probability of defects in both of these processes becomes

$$P(\overline{X}_{M1} \cdot \overline{X}_{M2}) \leq 10^{-9} \cdot 10^{-9} = 10^{-18}$$

The relation between process defects and their effects will now be investigated. Manufacturing has many complications, so this is a good point at which to address this issue, especially as the only remaining process is 'spurious mechanical damage' e.g. tool drops or transport mishaps.

21.2 GENERAL PROCESS DEFECTS AND EFFECTS

A general process can be described by the characteristics shown in Figure 21.1. The specifics are arrived at when the relation between defects and effects are determined during process characterization.

For example, the manufacturing process defect $\overline{P}_{IDi}$ in Process I, defect i has the effect $\overline{F}_{MI}$ and the probability of the combined event is

$$P(\overline{F}_{MI}\overline{P}_{IDi}) = P(V_{IL} \mid T_K \overline{X}_M \overline{P}_{IDi}) \cdot P(T_K \mid \overline{X}_M \overline{P}_{IDi})$$
$$\cdot P(\overline{X}_M \mid \overline{P}_{IDi}) \cdot P(\overline{P}_{IDi}) \tag{21.2}$$

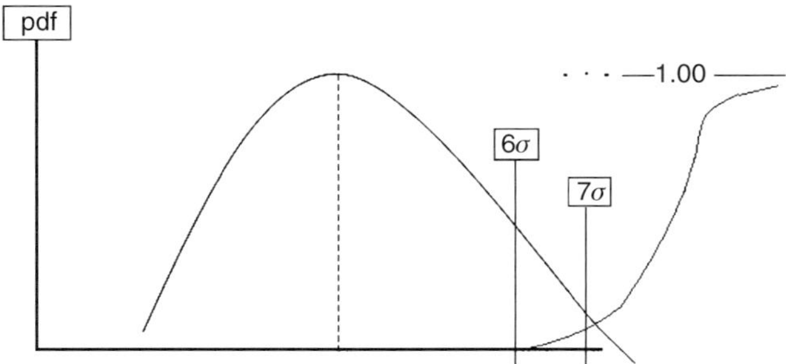

Figure 21.1 Process characteristics.

Equation (21.1) shows the probability of a certain reduction of a structural property due to a process defect i in process I. The first factor is the probability of the value of the reduction. The second is the probability of the type of property. The third is the probability of a certain process effect due to the defect. The defects considered for the process are:

- errors in timing
- errors in equipment
- errors in pressure
- errors in temperature
- errors in moisture.

The following expressions show the totality of the relation between defects and effects (Equation 21.3). We start from the probability of an unsafe state and assume that every element of safety has two parts. Each of the elements has a process and an implementation component respectively, e.g. design D has two parts, process D_P and implementation D_E, resulting in

$$
\begin{aligned}
P(\bar{S}_S) = {} & P(\bar{D}_E \mid D_P MIOR) + P(\bar{D}_P) + P(\bar{M}_E \mid M_P IOR) + P(\bar{M}_P) \\
& + P(\bar{I}_E \mid I_P OR) + P(\bar{I}_P) + P(\bar{O}_E \mid O_P R) + P(\bar{O}_P) \\
& + P(\bar{R}_E \mid R_P) + P(\bar{R}_P)
\end{aligned}
\tag{21.3}
$$

This results in the problems 'mishandling a good process' or 'having a bad process' and allows us to suspect either the process itself or the implementation of the process.

Equation (21.2) can be expanded further to reach a relation between, for example, the probability of a reduction of a specific property and a specific defect in a manufacturing. The following is the result:

$$
\begin{aligned}
P(\bar{F}_{MI}\bar{P}_{IDi}) = {} & P(V_{KL} \mid T_L \bar{X}_M D_1 T_K \bar{P}_Q) \cdot P(T_L \mid \bar{X}_M D_1 T_K \bar{P}_Q) \\
& \cdot P(D_1 \mid T_K \bar{X}_M \bar{P}_Q) \cdot P(T_K \mid \bar{X}_M \bar{P}_Q) \cdot P(\bar{X}_M \mid \bar{P}_Q) \cdot P(\bar{P}_Q)
\end{aligned}
\tag{21.4}
$$

Here:

The first factor on the right-hand side is the conditional probability for the value of the reduction
The second factor is the conditional probability of the type of property
The third factor is the conditional probability of the extent of the defect
The fourth factor is the conditional probability of the type of the defect
The fifth factor is the conditional probability of the effect
The sixth factor is the probability of the process quality.

21.3 MAINTENANCE DEFECTS IN SAFETY MANAGEMENT

The following defects in processing, i.e.

- defects in scheduled maintenance;
- defects in inspection;
- defects in repairs; and
- spurious mechanical damage.

can have the following effects:

- increase in loads due to wear and tear;
- large damage remaining due to non-detection;
- reduced strength due to structural property reduction;
- increase in loads due to faulty geometry; and
- mechanical damage included in repairs.

21.4 OPERATIONAL DEFECTS

These kinds of defect can include the following mishaps:

- higher than limit load;
- exposure to severe environments;
- walk-around inspection 'non-detection';
- non-detection and non-reporting of ground damage.

They can have the following effects:

- short-term limit exceedance;
- short-term load exceedance or damage due to reduced structural properties;
- undetected, potentially serious damage (extensive internal damage);
- unreported large internal damage.

21.5 DEFICIENT REQUIREMENTS

These can include the following in terms of misleading defects:

- metal-based requirements;
- damage tolerance size requirements and detectibility;

- damage resistance minimum requirements for fail-safe and discrete damage;
- B-value requirement and need for fail-safety.

They can have the following effects:

- irrelevant design drivers;
- unsafe damage tolerance design;
- damage resistance validation a case-by-case need;
- mean value allowable values instead of B-values unsafe.

21.6　DEFECT AND EFFECT REVIEW

This review of defects and effects brings us back to the probability of an unsafe state, Equation (21.3), and the relation between defects and effects (Equation 21.4).

The more detailed version of the equation of the probability of an unsafe state has each element of safety split into process and application components. Figure 21.2 illustrates the nature of the process.

21.7　STRUCTURAL SAFETY

Equation (21.3) describes the influence of all elements of safety on the probability of an unsafe state. The equation describes the contribution at every DDP for all PSEs. The model of safety of the total structure starts at the DDP level and sums over all DDPs and PSEs.

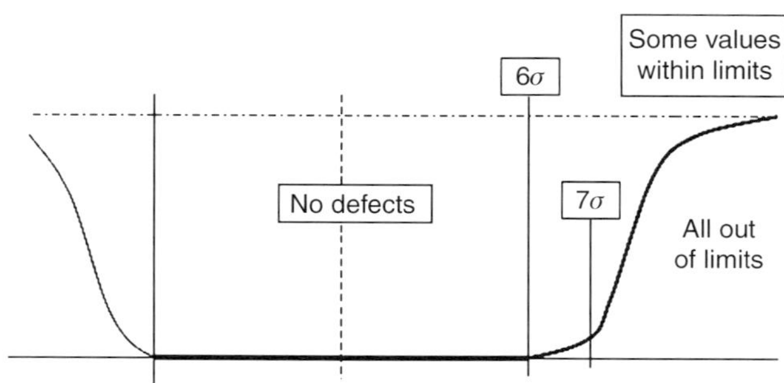

Figure 21.2 Process definition.

The analysis of a DDP involves both the reliability of the application of the process and the quality of the process. The following requirements apply to the process at every DDP. The equation is as follows:

$$P(\bar{S}_S) = P(\bar{D}_E \mid D_P MIOR) + P(\bar{D}_P) + P(\bar{M}_E \mid M_P IOR) + P(\bar{M}_P)$$
$$+ P(\bar{I}_E \mid I_p OR) + P(\bar{I}_P) + P(\bar{O}_E \mid O_P R) + P(\bar{O}_P)$$
$$+ P(\bar{R}_E \mid R_P) + P(\bar{R}_P) \tag{21.5}$$

The above content addresses two different kinds of unsafe state. The first contribution for each element is the unsafe execution of the element, given that all other elements are safe, the second item being an unsafe process. For example an unsafe process could have a criticality determination that uses ultimate allowable values based on a definition of state of damage, state of the environment, disturbance in internal loads due to fasteners and specific loads. The competing load cases could involve (at load reversal):

- compression with damage, fastener effects and an environment compatible with the load-case in question;
- tension with damage, fastener effects and an environment compatible with the load-case in question;
- shear with damage, fastener effects and an environment compatible with the load-case in question;
- buckling with damage, fastener effects and an environment compatible with the load-case in question;
- crippling with damage, fastener effects and an environment compatible with the load-case in question.

The internal loads are typically biaxial with shear and require conclusions as to failure criteria and buckling interaction to determine criticality.

Process definition is based on the objective that defects are outside the 6σ interval and the defects are distributed between 6σ and 8σ as shown in Figure 21.2, with modestly, changed ($<20\%$) defects at between 6σ and 7σ.

The evolved version of the equation for the probability of an unsafe state includes five terms that describe the part of an unsafe state that is associated with the processes intended to improve the level of safety. The purpose of these processes is to provide the guidance required to produce safe airplanes when new materials, new processes and new structural concepts have been developed and matured processes are not in place. The five types of processes and their specific element of safety are:

- design processes
- manufacturing processes

- maintenance processes
- operational processes
- requirements development processes (composites regulation).

The development and 'shakedown' of these processes are an extremely important part of any airplane development program and the foundation of safety management. Equation (21.5) clearly reveals that with the terms

$$P(\bar{D}_P) = P(\bar{M}_P) = P(\bar{I}_P) = P(\bar{O}_P) = P(\bar{R}_P) \approx 0 \tag{21.6}$$

all the processes are mature and do not contribute to an unsafe state, but make all processes safer.

It is interesting to note that a well-educated development team with 'practicing engineering know-how' is a firm foundation for safe processes, and that the processes satisfy Equation (21.6).

Hence the process development results in the following operative equation for the probability of an unsafe state:

$$P(\bar{S}_S) = P(\bar{D}_E \mid D_P MIOR) + P(\bar{M}_E \mid M_P IOR) + P(\bar{I}_E \mid I_P OR)$$
$$+ P(\bar{O}_E \mid O_P R) + P(\bar{R}_R \mid R_P) \tag{21.7}$$

On the right-hand side of the above equation:

The first term describes the probability of unsafe design, given safe design process, safe manufacturing, safe maintenance, safe operation and safe requirements

The second, the probability of unsafe manufacturing, given safe manufacturing processes, safe maintenance, safe operation and safe requirements

The third, the probability of unsafe maintenance, given safe maintenance processes, safe operation and safe requirements

The fourth, the probability of unsafe operation, given safe operational processes and safe requirements

The fifth, the probability of unsafe requirements, given a safe requirements formulation process.

Equation (21.5) contains five terms describing the probability of an unsafe state and is an interesting version thereof. The terms are:

$P(\bar{D}_P)$: probability of an unsafe design process
$P(\bar{M}_P)$: probability of a set of unsafe manufacturing processes
$P(\bar{I}_P)$: probability of a set of unsafe maintenance processes
$P(\bar{O}_P)$: probability of a set of unsafe operational processes
$P(\bar{R}_P)$: probability of an unsafe requirements formulation process.

Equation (21.5) shows results in relation to reliable, mature processes. The usual situation is that the design process is unreliable, often being based on 'metal thinking'. The requirement formulation process lacks empirical support and often relies on metal-based rules of thumb, and the manufacturing processes too often produce defects within the 'unsafe' range.

The design process is far from safe, $P(\overline{D}_P) \neq 0$, and this will not cease to be the case until consistent, and reliable process is put in place (failure criteria, buckling inter-action, buckling with damage, crippling with damage, design data scale-up, etc.).

The requirement formulation process is essentially absent, $P(R_P) \approx 0$, and it will not be in the short term before and a safe process will be in place (damage tolerance strength, fail-safe loads, damage rules, criticality, etc.)

Chapter 22
Unsafe State

Equation (21.7) describes the probability of an unsafe state in relation to the effects of all of the elements of safety. The following terms contribute:

> The first term: the probability of an unsafe design given a safe design process, safe manufacturing, safe maintenance, safe operation and safe requirements;
>
> The second term: the probability of unsafe manufacturing given safe manufacturing processes, safe maintenance, safe operation and safe requirements;
>
> The third term: the probability of unsafe maintenance given safe maintenance processes, safe operation and safe requirements;
>
> The fourth term: the probability of unsafe operation given safe, operational processes and safe requirements;
>
> The fifth term: the probability of unsafe requirements given safe requirements formulation process.

22.1 UNSAFE ELEMENTS OF SAFETY

The unsafe elements involve moderate defects in the processes that are of the order of magnitude of ultimate margin of safety 20% pertinent reduction of properties or increases in loads or spurious mechanical damage characterized as between 7σ and 8σ.

22.1.1 *Unsafe design*

The main threat is mechanical damage that violates fail-safe integrity and discrete source 'get-home load integrity' or increased internal load levels ($P < 10^{-11}$). The initial objective of securing a process quality of 6σ has shown potential, and if we start to investigate the range associated with the unsafe state we find that for a permanent reduction in strength, stiffness and toughness of 15% or a permanent increase in internal loads the limiting probability is 10^{-12}.

22.1.2 *Unsafe manufacturing*

The main threat is widespread damage resistance and damage tolerance reduction violating discrete source damage. We now look at Example 22.1, which deals with reduction of strength.

Example 22.1 We assume a 15% strength reduction and the resulting probability of survival $1 - 0.023 = 0.977$; for n flights we have the following results:

n flights	Probability of surviving n flights
1	0.977
10	0.10
1000	10^{-11}

And for 15% permanently increased internal loads we have

n flights	Probability of surviving n flights
1	0.991
100	0.40
1000	10^{-4}

the results in both tables corresponding to nominal defects.

22.1.3   *Unsafe maintenance*

A similar situation to reduced structural properties caused by defects in processing and increased internal loads in both repair and assembly or by defects in scheduled mainte-nance will lead to nominally acceptable defects.

22.1.4   *Unsafe operation*

Defects in operation can result in damage, or accidental damage can fail to be detected, which reduces structural properties and thereby brings about an unsafe state at 7σ.

22.1.5   *Unsafe formulation in requirements*

Unsafe interpretation of strength requirements or an increase in internal loads due to unconsidered defects in geometry and quality will also contribute to a 7σ probability of an unsafe state.

22.2   TOTAL PROBABILITY OF AN UNSAFE STATE

The total probability of an unsafe state for a detail design point (DDP) is

$$P(\bar{S}_S) = 10^{-11} + 4 \cdot 10^{-12} = 1.4 \cdot 10^{-11}$$

which corresponds to a principal structural element (PSE) value, if there are 50 DDPs, of $0.7 \cdot 10^{-9}$ and for 100 PSEs, $0.7 \cdot 10^{-7}$, which clearly satisfies the present requirement, but for 100 DDPs it becomes $1.4 \cdot 10^{-9}$ and for 100 PSE, $1.4 \cdot 10^{-7}$. However, the required safety value as projected for the anticipated growth would not and could not be satisfied without new materials, inspection techniques and improved quality control.

An effective long-term defects distribution will have to include a transition zone with a gradually increasing defects size from zero to 10% at between 6σ and 7σ.

$$P(\overline{S}_S) = 1.2 \cdot 10^{-11}$$

More details can be found in Chapter 24.

Chapter 23
Summary of Safety Management

Safety management would assure safe composite structures by using high-quality processes in the production of transport-category airplanes, recognizing that damage and defects have to be either avoided or detected, making inspection and quality control primary defenders of safety.

The primary protection against 'undetected loss of integrity' is provided by making the events either very rare or very easy to detect. Hence the natural complement to quality control and major inspections is a monitoring – updating process in service that updates probabilities of unsafe states in time and reduces risk and uncertainty by adjusting inspection periods and inspection approaches.

Before- and in-service elements of safety are considered for composite structure.

- structural design
- manufacturing
- maintenance
- operation
- requirements formulation.

Each element consists of two parts, process and application. In service, a monitoring process analyzes service and inspection data to update probabilities and to update and correct safety levels, by managing risk and reducing uncertainty, and adjusting inspection periods and approaches adopted.

The design works with the probability of an unsafe state, making damage sizes, probabilities and detection probabilities part of the design process.

The elements of safety are required to have a 6σ process quality (no defects within the interval), and to ensure that the start of the unsafe region lies at 7σ. The approach to be followed deals with what existing transport-category technology has demonstrated. What is required for evolving composite structure and what is necessary over the next 20 years to avoid a rise in fatalities?

The conclusions are that the 6σ quality requirements for the processes involved must satisfy the general requirements for the probability of an unsafe state, and that design must satisfy consistent use of B-values and fail-safety.

Future safety challenges include widespread defects affecting structural properties, a design approach to discrete source damage and get-home loads, damage resistance and determination of maximum damage growth rates, and the development of a design process for safety, criticality and reliable fail-safety.

23.1 DESIGN PROCESS AND PRACTICES

Safety-dependent actions can require adjustment to meet safety levels that have been deemed part of today's requirements and in some cases extrapolated into the future. It is useful to remember that the current requirement of one unsafe flight in 100 000 flights (structural share 10%) was accepted by ex-President Clinton's Commission on Security and Safety and confirmed this year (2007) by the FAA. The target of another 10% improvement over the next few decades, to one unsafe flight in one million flights, has also been recommended.

The safety evaluation has recommended that present practices in design need to be supported by a design process that has been developed by the 'practicing' community and that is supported by the FAA.

23.1.1 Design criticality

Composite structural design has a set of ultimate, critical modes. It is in general based on internal loads that are biaxial with shear, requiring application of failure criteria and buckling interaction rules. In many locations in airplane structure internal load reversals occur, resulting in load cases that produce either compression or tension in a primary direction at DDPs. There are also many locations where load redistribution causes significant shear. Thus B-quality design values, taking their definition to be that usually found in everyday practice, can be written as

$$P(\bar{B}_U S_D S_E S_F S_L) = P(\bar{B}_U \mid S_D S_E S_F S_L) \cdot P(S_D \mid S_E S_F S_L)$$
$$\cdot P(S_F \mid S_E S_L) \cdot P(S_E \mid S_L) \cdot P(S_L)$$

Here the first factor on the right-hand side is the allowable value.

The second factor is the probability that the state of damage is compatible with the environment, fastener effect and the load.

The third factor is the probability of the fastener effects being compatible with the environment and load-case.

The fourth factor is the probability that the environment and load-case being compatible, can occur at the same time.

Thus the straight criticality could mean that each load-case and critical mode could have a different critical damage, that the environment, e.g. temperature, could be $-65F$, RT or T_{MAX}, whichever is most appropriate, and fastener correction could be different for different types of stress. This process would lend itself to software design and perhaps to being combined with failure criteria, so for example crippling has often been associated with very specific critical damage in the design.

23.1.2 *Widespread defects*

The widespread reduction of strength can result in loss of fail-safe integrity, a state which is unacceptable if damage tolerance does not turn out to be critical and provide an ultimate margin of safety that can be used as an extra reserve up to the value of the margin of safety. Therefore if the reduction turns out not to be larger than the margin of safety, then the situation would be acceptable.

$$MS = \frac{F_U}{f_U} - 1 \Rightarrow MS = 0 = \frac{kF_U}{f_U} - 1 \Rightarrow k = \frac{f_U}{F_U}$$

and if the limit damage is three times as large as the ultimate gauge then:

$$t_U = \frac{N_U}{F_U} \quad t_L = (1.5\sqrt{.33})^{-1} \frac{N_U}{F_U} = 1.16t_L \Rightarrow MS = +14\%$$

Damage tolerance design provides a 14% margin of safety.

23.1.3 *Discrete damage integrity*

Discrete source damage comes in many forms, for example bird strike in the nose or leading edges, birds of different size being taken into account in relation to design. Damage occurs due to loss of damage resistance and the resulting loss of 'get-home load' capability, having its source in manufacturing defects. The overall process is characterized by a detail design testing process that, from a safety standpoint, must be run in parallel with process development for the production of this type of critical structure.

The objective of this design is to avoid impact damage with penetration or loss of get-home load integrity or of the PSE, or direct failure. The manufacturing process involved if based on 6σ, would be naturally limited below 7σ to 8σ as this is the range in which defects begin to become serious, particularly when scatter is increasing.

We start with the combined loss of damage resistance and damage tolerance, which is governed by the equation

$$P(\bar{X}_M X_{DD} \bar{D}_{GH} \bar{B}_{GH} \bar{A}_{GH}) = P(\bar{A}_{GH} \mid \bar{B}_{GH} \bar{D}_{GH} \bar{X}_{DD} \bar{X}_M) \cdot P(\bar{B}_{GH} \mid \bar{D}_{GH} \bar{X}_{DD} \bar{X}_M)$$
$$\cdot P(\bar{D}_{GH} \mid \bar{X}_{DD} \bar{X}_M) \cdot P(\bar{X}_{DD} \mid \bar{X}_M) \cdot P(\bar{X}_M) \qquad (23.1)$$

Example 23.1 A numerical illustration of Equation (23.1) will now follow in an attempt to capture the nature of the incident.

The first factor on the right-hand side describes the Probability of failure, given discrete 'large' size initial damage.

The second factor, the Probability of loss of 'get-home' integrity, given 'large' initial damage and manufacturing defects (reducing damage resistance and damage tolerance).

The third factor, the probability of too-large 'Get-Home' load damage initially, given a discrete source damage and manufacturing defect event.

The fourth factor, the probability of a discrete source event, given manufacturing defects.

The fifth factor, the Probability of failure, given bird strike and manufacturing defect

$$P_F = 10^{-1} \cdot 10^{-1} \cdot 10^{-6} \cdot 10^{-9} = 10^{-17}$$

which is a conservative evaluation. This is a difficult task, because scatter often increases with reduction of most kind of strength reductions. A thorough evaluation is needed for the process characterization.

23.2 DISCRETE SOURCE DAMAGE DESIGN

A number of categories belong under this heading: bird strike, engine turbine blade impact, tire-burst, hailstone impact in flight, and up to $20\,\text{ft}^2$ breach in pressure confinement structure. All of these have empirical roots in the 'metal world', but take on new meaning in safety management.

Very little of the experience gained in the world of metal carries over to composites so the way to proceed involves development design testing, which naturally is often combined with process characterization for new materials and processes.

23.3 TIRE-BURST

Landing after an aborted take-off occasionally leads to tire-burst due to the heat generated from the energy accumulated in the gear brakes. There are two safety threats in the main wheel-well. Penetration of wing center section rear spar resulting in leaking fuel, and penetration of the pressure deck and disabled passengers. Both events involve penetration of structure due to exploding tires. The governing equation is:

$$P(\bar{P}_{TB}\bar{X}_{TB}\bar{X}_M\bar{T}_R) = P(\bar{P}_{TB} \mid \bar{T}_R\bar{X}_{TB}\bar{X}_M) \cdot P(\bar{X}_{TB} \mid \bar{T}_R\bar{X}_M) \cdot P(\bar{T}_R) \cdot P(\bar{X}_M) \qquad (23.2)$$

Here:

$\bar{P}_{TB}$ is penetration due to tire-burst
$\bar{X}_{TB}$ is the event tire-burst

manufacturing defect $\overline{X}_M$ is the causing reduced damage resistance

$\overline{T}_R$ is the event 'aborted take-off'.

Example 23.2 Given here a numerical illustration of Equation (23.2) with damage resistance defect:

$$\text{Probability of penetration} = 10^{-2} \cdot 10^{-2} \cdot 10^{-2} \cdot 10^{-9} = 10^{-15}$$

This shows the threat from tire-burst even when defects are not present, and even if only loss of damage resistance is involved. Testing that shows critical penetration size data is important, and not only when defects are present.

The safety requirements when defects are not present can be written as

$$P(\overline{P}_{TB}\overline{X}_{TB}\overline{T}_R\overline{X}_O) = P(\overline{P}_{TB} \mid \overline{X}_{TB}\overline{T}_R\overline{X}_O) \cdot P(\overline{X}_{TB} \mid \overline{T}_R\overline{X}_O) \cdot P(\overline{T}_R) \cdot P(\overline{X}_O) \qquad (23.3)$$

If this event is considered a defect, $\overline{X}_O$, in operational procedure, the following numerical result is arrived at:

The probability of penetration is $= 10^{-2} \cdot 10^{-2} \cdot 10^{-3} \cdot 10^{-9} = 10^{-16}$, if operation is a 6σ process.

23.4 FAIL-SAFE BACK-UP FOR REDUCED DAMAGE TOLERANCE DUE TO MANUFACTURING DEFECTS

This requires fail-safe design criteria for manufacturing defects causing loss of toughness and load-path failure at limit, external load and ultimate, internal load redistribution. This is particularly true for compression in composites, which can be a violent event and may have a dynamic load factor of up to 2.0. This is expected to be a test-validated situation with failure of a load-path under limit, internal loads and load redistribution. Because composites are significantly different from metals in terms of energy absorption, and because of safety management the resort to fail-safe design for back-up to defects that cause loss of damage resistance and fracture mechanics based damage tolerance due to manufacturing defects. A thorough test program must be used to determine the dynamic increase in design load. A fail-safe test program is a definite requirement for structural safety, and should be considered as part of composites regulation and design criteria.

23.5 HAILSTONE IMPACT IN FLIGHT

In the last few years there have been an increasing number of incidents involving hailstone impacts in flight causing severe damage to transport category airplanes. This has mainly involved metallic airplanes. However, emerging composites airplanes are so much more sensitive to impact damage that it is time to include this type of accident under the heading of discrete source damage. The planning of a detail design test program to produce a more damage-resistant surface structure and to determine the limits of the impact survivable could possibly be replaced by an operational responsibility to avoid these types of weather system. Requisite information and weather updates should be the operational responsibility of weather forecasting and weather reporting to avoid these types of event. A 6σ process could be implemented and the guiding equation could be expressed as:

$$P(\bar{D}_{HD}\bar{X}_{HD}\bar{U}_{HD}\bar{X}_O\bar{A}_{HD}) = P(\bar{A}_{HD}\mid\bar{U}_{HD}\bar{D}_{HD}\bar{X}_{HD}\bar{X}_O)\cdot P(\bar{U}_{HD}\mid\bar{D}_{HD}\bar{X}_{HD}\bar{X}_O)$$
$$\cdot P(\bar{D}_{HD}\mid\bar{X}_{HD}\bar{X}_O)\cdot P(\bar{X}_{HD}\mid\bar{X}_O)\cdot P(\bar{X}_O) \qquad (23.4)$$

A numerical assessment of the safety threat is presented in the next example.

Example 23.3 The following is an illustration of the safety level that might be acceptable. Probability of failure with desirable constraints $= 10^{-1}\cdot 10^{-2}\cdot 10^{-3}\cdot 10^{-9} = 10^{-15}$. A stringent 6σ communications process could make this a very improbable event.

A detail design development testing program with the object of determining the critical size of impact damage at pertinent locations could become the bulwark of safety.

23.6 PROCESS DEVELOPMENT

Six-Sigma process quality development has the empirical long-term result that the 6σ zone is defect-free. The defects are therefore distributed from 6σ upward.

Most of the structural property data have a coefficient of variation in the range 0.1. Thus, the range of the defects starts at 6σ and at the end of the first sigma-interval will be of the order of magnitude $\sigma = C_V\mu = 0.1\mu$, which corresponds to a probability of 10^{-12}.

$$\Phi\left(\frac{USR - 1.035USR}{C_V 1.035USR}\right) = \Phi\left(-\frac{0.034}{0.1}\right) = 0.34$$

The probability of the property, reduced by 10%, is 0.34. This can be considered an acceptable value especially with an ultimate margin of safety arrived at due to damage tolerance criticality.

The main reason for recommending 6σ processes, even though process development is part of the initial design, is that it sets the foundation for the design and has flexibility for continued improvement. The unbiased application of safety management to the design process requires an extensive insight into process characteristics, giving a total picture of the manufacturing processes, in particular defining the relations between cause and effect so that sound processes are developed and the 6σ region is defects-free and engine burst is resolved by blast containment instead of structural redesign.

The use of materials and processes with no in-service experience requires extensive characterization, improvements (defined cause and effects between defects and structural inadequacies), so that the total safety threat produced by the elements of safety can be assessed at the beginning of the design stage.

This book has taken the recommendations of ex-President Clinton's Commission of Security and Safety in Aviation as the basis for the safety level that should be met by transport-category airplanes. The resulting requirements were confirmed by the FAA earlier this year (2007). It also spelled out the safety targets for the next two decades.

The results of the safety management study presented in this book stress that the occurrence and detection of damage in composite structure are crucial issues. The nature of selected and characterized processes with minimum defects must be available early in the design phase to avoid nasty surprises. An understanding of the extent of multiple effects due to a single defect, and findings for the 'next system', may be in conflict with these findings.

Situations wherein process defects have deleterious effects on, for example, strength, stiffness and toughness, can cause cost overruns and schedule slides and be very difficult to resolve. The message is that concurrent process development and design development is necessary for safe structural design. We cannot afford surprises!

The fact is that the totality of the safety problem cannot be addressed practically until defect types and couplings have been determined and underestood, and the defects have been pushed out of the 6σ interval. The time to carry out the detail design of structure is when data and measures of structural safety are available.

This investigation is based on safety requirements that can be traced back to the requirements of structural safety for 'one unsafe flight in one million', shared between design, manufacturing, maintenance, operation and requirements formulation, and supported by a monitoring process that preserves safety levels and reduces uncertainty.

The processes identified and their quality and maturity become central issues in structural safety. The elements of safety discussed have a shared responsibility in establishing, maintaining and restoring levels of safety.

The selection of processes and continued quality improvement of these processes are paramount to safety. The elements of safety investigated in this book are:

- structural design
- manufacturing
- maintenance

- operation
- requirements formulation
  and
- in-service monitoring.

Each has two parts – process and utilization – and all have a direct influence on levels of safety.

New materials and new structural concepts must have fully developed processes before safety and cost impacts can be assessed. The advantage in aiming for 6σ quality is the flexibility and long-term defects probabilities that appear to make safety objectives attainable, once the nature of the defects and their combination is known.

The total set of challenges to meet the quoted safety requirements involves the use of B-value panel allowable values, B-value residual strength, the probability of limit damage sizes, and probability of detection (while quality control still remains an expensive way to achieve safety and 6σ process qualities).

The focus on 6σ process quality consists of the successful developmental efforts undertaken in many parts of the industry. Successful implementation falls within the safety targets identified, an extensive application description selection set being available; the initial targets are within the range of the objectives, but extensive material characterization is required to initiate 6σ developments, which all are data-based.

The following cannot be said often enough: 'Safety management, the roles of elements of safety, the total design in relation to safety constraints, the service monitoring and high quality process development merge to form a total safety- and quality-coordinated effort where cost is uppermost on the scale of priorities'.

The integrated or coordinated joint activities that commence with totally focused development activities among the elements of safety start with process development and continue with safety management; this includes design process development and integration of requirements formulation and the establishment of validated practices, along with coordinating service monitoring.

23.7 CONCLUSIONS

The safety management process requires that high-quality processes are employed resulting in 6σ processes with a transition zone in defect size between 6σ and 7σ.

Discrete source damage requires design process development-based point design testing.

A back-up fail-safe criterion requires fail-safety under applied limit external loads.

Chapter 24
Philosophy – Design to Explicit Safety Constraints

Safety management places great emphasis on process development. The selection of 6σ process quality puts more emphasis on specifications and, especially when design requirements, design data and allowable values enter into the specifications, we find that realism must be the only operative requirement. The author reminds you of the classic compromise situation for composites that has existed over the last couple of decades. We have heard, *ad nauseam*, about the contradictory requirements between 'open-hole compression' and 'compression after impact' without discovering a single open hole in the structures involved. There is much structure suffering impact where the prevailing temperature is $-65F$, while allowable values are measured at room temperature, a balance which is directly driving the results in process development that distinguishes between probabilities of 10^{-9} and 0.0000034. The requirements include damage resistance, damage tolerance and maximum damage growth rates and, as shown in the previous chapter, practical standards are still to be adopted in the practicing community.

Discrete source damage has a far from a consistent set of requirements, and an informed set of needs for process development, e.g. damage-containment-based 'engine burst' design through containment, still has to be developed. Without this, safety management has a long way to go before it finds a productive way to achieve safety. But, at least, the participating players have been identified.

Both coupling between a particular defect and a number of particular structural properties and requirements for internal loads increases the need for a consolidated, competitive objective for the process requirements.

The coupling between material selection and process development can be dealt with as soon as a consistent set of targets has been introduced into process development. If one wants to develop a practical set of design data, all the right compromises need to be identified in a way that has been mentioned earlier:

- compression strength, state of damage, state of environment, state of local load concentration, global state of internal loads;
- buckling, state of damage, state of environment, state of local load concentration, global state of internal loads;
- crippling, state of damage, state of environment, state of local load concentration, global state of internal loads;
- tension strength, state of damage, state of environment, state of local load concentration, global state of internal loads;

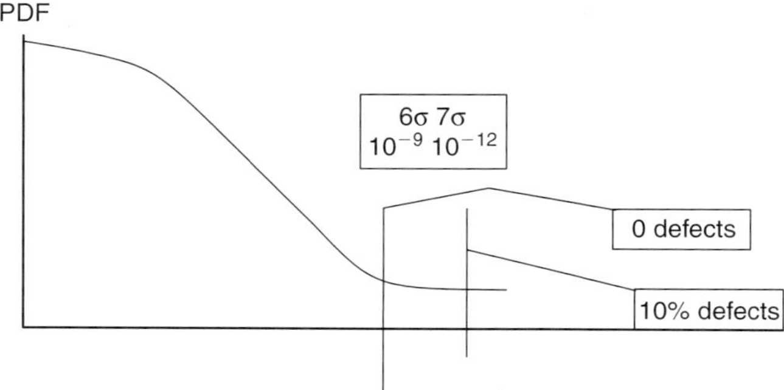

Figure 24.1 Six-Sigma limits.

- shear strength, state of damage, state of environment, state of local load concentration, global state of internal loads.

Figure 24.1 describes the 6σ boundary and the 7σ point where the defect shows 10% reduction of allowable quality. The allowable data, once defect-free, are

$$C_V = \frac{\sigma}{\mu} \Rightarrow \sigma = C_V \mu \quad \text{and} \quad C_V = 0.1 \Rightarrow \sigma = 0.1\mu$$

Thus, if we have 10% reduction at 7σ we have, for the probability of the event

$$P(\bar{U}P_{7\sigma}) = P(\bar{U} \mid P_{1\sigma}) \cdot P(P_{1\sigma})$$

and

$$P(\bar{U} \mid P_{7\sigma}) = \Phi\left(\frac{F_B - 0.9\mu}{0.1\mu}\right) = \Phi(8.7 - 9.0) = 0.5$$

and

$$P(\bar{U}P_{1\sigma}) = 0.5 \cdot 10^{-12}$$

Here:

$\bar{U}$: 'loss of integrity'
$P_{7\sigma}$: process at 7σ.

So, the probability of an unsafe state from the last example becomes

$$P(\bar{S}_S) = 10^{-11} + 4 \cdot 0.5 \cdot 10^{-12} = 1.2 \cdot 10^{-11} \tag{24.1}$$

In this example we find that structural design is critical in damage tolerance and that the other four elements of safety contribute an additional 20% in weight when the defects are widespread. In this situation the process is in a competitive order of magnitude with the probability of an unsafe state and the number of processes per PSE becomes the deciding factor. So we are back to the importance of getting causes (defects) and effects characterized, and success in finding a defect-free 6σ interval, and the actual 6σ process developments, are for the future to tell because this is a very onerous undertaking even though many positive results have been claimed in total-quality-management circles. The number of processes included, and their individual successes, is where the final word will be said, but the enterprise is no doubt worthwhile in terms of demonstrated safety, especially bearing in mind the environment of uncertainty that surrounds composites.

It is worth mentioning that the most promising results claimed come with the 6σ quality, and there are no competing results shown for any alternatives in process development to meet the safety objectives derived for safety management and thus far deemed necessary by the FAA.

24.1 PROBABILITY OF FAILURE

The total probability of an unsafe state can be shown to converge to a probability of failure between major inspections provided loss of integrity occurs. It can be written at a detail design point (DDP) as

$$P(\overline{S}_S) = 10^{-11} + N_P 0.5 \cdot 10^{-12} \tag{24.2}$$

The probability of an unsafe state is shown below for each principal structural element (PSE), the total structure being shown.

$$P(\overline{S}_S) = \sum_{j=1}^{N_P} \left(\sum_{i=1}^{N_j} 10^{-11} + N_{Pi} \cdot 0.5 \cdot 10^{-12} \right) \tag{24.3}$$

Here, the number of processes at the DDP is N_{Pi}. N_i is the number of DDPs at each PSE.

Example: number of PSEs $= 100$; number of DDPs $= 25$; number of processes $= 10$. So the probability of an unsafe state for the structure is

$$P(\overline{S}_S) = 2 \cdot 10^{-11} \cdot 25 \cdot 100 = 0.5 \cdot 10^{-7}$$

As is shown in Equation (24.2) there are two safety components, some of the contributions not being obvious. Some of the less apparent include the following three processes:

1. design process
2. requirements formulation
3. in-service monitoring, analysis and updating.

24.2   DESIGN PROCESS

The design of new transport-category composite airplane structures is far from routine. A comparison between existing practices and existing design processes immediately brings up the fact that there are no existing FAR regulations for composite structure in existence. If there were, they would drive the development of practices in the composites commercial community and, as the military tends to use a different class of material and to some degree different requirements, this quest to develop a new design process should be coupled with the evolution of a process that deals with requirements formulation (FAR 25) regulations, criteria and all pertinent guidance material.

Existing FAR 25 regulations mainly deal with the rules of the 'metal world'. However, in actuality there is no real support for composites design; rather, it is the other way around. Existing design processes are, in fact, totally influenced by metal-related ideas and will continue to be so until a new process developing composites requirements for composite structures is in place and working, so that what is typical for composites is recognized and incorporated in regulations and practices.

We need to recognize and accept that composite structure often is critical to damage. That means, for example, that there are positive ultimate margins of safety and that discrete source damage scenarios might be critical. It might mean that the fail-safe internal loads redistribution are the critical limit event, or, possibly, that damage resistance might be the critical property in laminates, or bonded or co-bonded composite structures.

Ultimate allowable values are characterized by the statistical requirements for A-values, B-values, etc., state of damage, state of environment (e.g. temperature compatible with load case), state of local foci of internal loads due to fasteners, or the definition of global load, which for composites creates many more combinations than what normally is the case in the 'metal world'. Load reversal creates more complications in composites. Failure criteria and buckling interactions are far more complicated in composites. So if you look at the true margins of safety in composite limit-critical structure and add to that the effects of processing defects, you begin to discern some of the basic differences. An added fact is that critical damage sizes are not spelled out in any existing regulations, and there is no predominant praxis in use.

The concept of processing defects, and what their effects are, add yet another complication to safe structure, and must be considered as part of the safety evaluation.

In the final evaluation of how safe structure is, this matters, and not everything is caught in quality control. The design process for each PSE must contain considerations of the effects of the process defects.

Example 24.1　An evaluation of the sizing algorithm for *widespread defects* when the probability of an unsafe state is based on a detail design point (DDP) with four processes involved gives:

$$P(\overline{S}_S) = 1.2 \cdot 10^{-11}$$

This equation is satisfied when the residual strength for limit damage size is equal to the design B-value for the PSE with a 20% margin of safety.

The result is based on the overarching requirement for structural probability of failure of 10^{-6} or less and 6σ process quality. Clearly the safety management has a 20% penalty for a safe airplane.

24.3　REQUIREMENTS FORMULATION

The development of safe design practices for composites depends for its success on the evolution of processes of regulations for composites for transport-category airplanes that address:

- damage tolerance
- damage resistance
- maximum damage growth
- ultimate allowable values, and definitions
- failure criteria
- buckling interaction
- process for criticalities
- discrete source damage (design development testing)
- buckling and crippling with damage
- fail-safe design and testing.

This initial list identifies many failures and requires a technical approach where objective arguments and technical demonstrations prevail; not an election or political inputs but the engineering arguments presented by the practicing community can assure a 'late' success. Furthermore, a coordinating community is a good thing, but the categories themselves should be kept separate, because the actual interests within the different coteries are different and rarely objective.

Without change in process for realistic regulations there will be no 'engineering processes' for the design of different categories of flight vehicles, to the detriment of both aviation and composites.

24.4 IN-SERVICE MONITORING

Uncertainty is a dominant state of mind in the composites world. Design of structures is not controlled by a well-understood process in the 'practicing' community. The experts have not yet 'landed' in this community. A large number of design decisions are not made as they are dominated by random variables and consequently have not been revisited.

Service data and inspection data, if properly monitored and analyzed in service, can contribute added knowledge both short-term and long-term, if supported by risk management and uncertainty reduction in service by updating databases for safety level control and updating of *a priori* understanding of random phenomena such as damage size distribution in service, detection probabilities and defects distribution, thus developing an environment of ever-increasing insights.

24.5 PROCESS DEVELOPMENT

Safety management, the practicality and the success of eliminating defects from the 6σ interval and achieving a gradual transition to 'gentle defects' in the 6σ to 7σ region, and the incorporation of all the processes of the elements of safety into the structural level of safety, are the fundamentals for attaining realistic and practical designs with moderate safety levels.

So, while requirements are enforced top-down, the design is conducted bottom-up. In the process of designing to safety constraints, the recent example illustrating order of magnitude of total safety levels of the five elements shows that it could be as low as 1.2 times the safety requirement of structural design. The idea of an equal share for all the elements would lead to 5.

The price for that development is not knowing the designed-in safety level until the process development has been completed. The total safety requirement will therefore depend on the complexity of the structure, the number of PSEs and the number of DDPs, which is not surprising in itself.

The supporting processes:

- design process
- requirements formulation process – regulations, criteria and practices
- monitoring process

must be put in place during the development of the composite airplane including all *a priori* distributions. However, it must be subject to continuous quality management, the objective of which will be to reduce risk, contain uncertainty, improve knowledge and gather service experience.

These processes, as well as the traditional processes, are not mature until compliance has been demonstrated through performance, which may have a continuous element protected through the monitoring process and updating of probability distributions that includes gathering data on defect severity and frequency, updating process reliability and reducing 'quality control corrections'.

Thus the validation of both 6σ and 7σ defect data will be the measure of maturity of the developed processes. The supporting processes, all of which must have an objective foundation, should also be part of this validation.

24.6 CONCLUSIONS

The development of process development for design and requirements formulation are essential for safety management.

A maturing (long-term) quality level is an essential part of process development.

A process is necessary for design of composite structure for discrete source damage.

Chapter 25
Elements of Safety and Structural Integrity

The elements of safety provide threats against different types of major structural integrity. The participating integrities are:

- ultimate integrity
- fail-safe integrity
- damage tolerance integrity
- discrete source integrity
- damage resistance integrity.

25.1 ULTIMATE INTEGRITY

The main purpose of ultimate integrity is, in most cases, to assure a factor of safety of 1.5 with limit external loads applied to the structure. The ultimate B-values are used for the primary structure providing fail-safe integrity with one load-path failed. The critical situation is widespread defects and lost fail-safe integrity (back-up to damage tolerance) and a very low number of flights of survival, if a positive ultimate margin of safety and accurate fail-safe analyses and design are not provided. The threats come in the form of the effects of defects from, for example, all of the elements of safety and accidental and mechanical damage of ultimate damage size.

25.2 DAMAGE TOLERANCE INTEGRITY

The threats that arise are effects of defects from the elements of safety and mechanical limit damage size, in addition to limit internal loads. B-quality limit allowable values are needed. This type of integrity is the most critical in many cases and is the main contributor to unsafe states in relation to the combination of limit damage size and the effects of defects.

25.3 DISCRETE SOURCE INTEGRITY

Discrete source integrity and fail-safe integrity are dependent on damage resistance and damage tolerance. Fail-safe needs internal load redistribution and moderate damage size to survive ultimate, internal loads (and limit external loads), especially in compression.

Discrete source impact needs to have enough damage resistance to survive the initial impact, and enough damage tolerance to survive 'get-home loads'. The design is expected to be conducted as a point-design testing program, which, for new materials, is the only option, even though complicated and expensive. Here, there is room for structural development.

It is worthwhile accepting that the different kinds of discrete source damage design have some features in common as regards the optimum design. The discrete source designs include:

1. bird strike
2. engine burst
3. tire burst
4. hail impact in flight
5. up to $20\,\mathrm{FT}^2$ pressure blow-out.

These threats are from a variety of sources and all subjective approaches are catastrophic; many types of defects can be involved. All of the different categories of integrity are involved:

1. Bird-strike impact can damage systems and injure crew.
2. Engine burst should be handled through containment.
3. Tire burst fragments can perforate fuel tanks, leading to injury of passengers and crew.
4. Hail impact in flight should be avoided (or new regulations should be applied).
5. General fail-safe capability should be provided, and compliance demonstrated through point design testing.

Design should be developed through point design testing to establish a firm foundation for service experience.

Appendix I
Undetected Loss of Integrity or Rare Loss of Integrity

The types of events discussed are natural events in the pursuit of safety because of the rarity of defects. In the world of composites the combination of ease of detection and good residual strength with damage create two powerful criteria.

The probability of undetected loss of limit integrity combined with failure can be written as

$$P(\bar{A}_n \bar{U}\bar{H}) = P(\bar{A}_n \mid \bar{H}_o \overline{UH}) \cdot P(\bar{H}_0) \cdot P(\bar{H} \mid \bar{U}) \cdot P(\bar{U}) \quad \text{which, when `}n\text{'} \to \infty$$

$$= P(\bar{H}_o) \cdot P(\bar{H} \mid \bar{U}) \cdot P(\bar{U}) = 10^{-2} \cdot 10^{-3} \cdot (10^{-1} \cdot 10^{-2} \cdot 10^{-3}) = 10^{-11}$$

Here the first factor is the marginal probability of non-detection
The second factor is the probability of non-detection of limit damage size

$$P(\bar{U}) = P(\bar{B}_L \mid \bar{X}D_5) \cdot P(D_5 \mid \bar{X}) \cdot P(\bar{X}) = 10^{-1} \cdot 10^{-3} \cdot 10^{-2} = 10^{-6}$$

The rarity of defects of the other elements of safety can be written as

$$P(\bar{A}_n \overline{UX}_E \bar{H}) \quad \text{when } n \to \infty = 1 \cdot 0.5 \cdot 10^{-12} \cdot 1$$

Here the probability of failure after the number becomes large is equal to 1; quality control detection is not a factor and does not adversely affect the outlook relating to cost, but is not a concern because of the rarity of an event.

The values for probabilities of survival in 'n' flights are shown in the table:

n flights	Probability of survival in 'n' flights $(P_{SU})^n$		
1	0.99	0.995	0.999
10	0.90	0.95	0.99
100	0.37	0.61	0.90
1000	$\sim 10^{-5}$	0.007	0.37
3000	10^{-14}	10^{-7}	0.05

This table shows that a failure in detection can be a serious problem for reasonable design missions, and the example of total probability of an unsafe state for a detail design point (DDP) becomes

$$P(\bar{S}_S) = 1.2 \cdot 10^{-11}$$

276

as a result of definitions on the previous page. A comparison for normal distribution probability of failure and B-value and mean-value design data, will found in Example AI.1 below.

Example AI.1 One of the difficult safety threats for composite-built transports is survival for at least the length of an inspection period with a damage size that gives the structure a limit residual strength capability. The situation requires a design mission during the period. We start by using B-value residual strength and damage tolerance criticality. We assume normal variables for strength and internal loads, and a comparison with mean-value residual strength allowable:

$$\text{The design mission has the following parameters: } \mu_y = -0.72\,LLR$$

$$\text{The strength has the following mean: } \mu_x = 1.15\,LLR$$

$$\text{Failure distribution mean:}$$

$$\mu_g = 0.43LLR \quad \text{with} \quad C_V = 0.1\,\sigma_g = (1.32 + 0.51)^{0.5}0.1\,LLR = 0.135\,LLR$$

yielding probability of failure $\Phi(-3.18) = 0.0008$ and probability of survival in n flights 0.9992.

This yields for design mission and mean value strength:

$$\mu_g = 0.28LLR \quad \text{and} \quad \sigma_g = 0.123\,LLR$$

yielding probability of failure $\Phi(-2.28) = 0.0113$, and probability of survival in n flights 0.9887:

Design mission and B-value		Design mission and mean-value	
n	Survival for n flights	n	survival for n flights
1	0.9992	1	0.9887
500	0.67	500	0.003
1000	0.45	1000	10^{-5}
3000	0.09	3000	10^{-15}

Clearly these results show several reasons why the 'mean allowable value' for residual strength is not acceptable. The probability of the B-value case is:

$$P(A_{SU}^n \overline{S}_L) = P(A_{SU}^{3000} \mid \overline{S}_L) \cdot P(\overline{S}_L)$$

This describes the combined event 'survival for 3000 flights with an unsafe state for limit damage'. The following example illustrates the probability of the defined, combined event:

Example AI.2   The first factor on the right-hand side of the above equation is 0.09 and the combined probability is

$$P(A_{SU}^{3000}\overline{S}_L) = 0.09 \cdot 10^{-11} \approx 10^{-12}$$

This value is acceptable considering the double indemnity nature of the combined event.

Hence it can be seen that B-value residual strength satisfies the safety requirements, but mean-value does not.

Appendix II
Sizing to Safety Constraints – Damage Tolerance Critical Structure

At B-value residual strength, when we satisfy the requirement that the probability of an unsafe state is $P(\bar{S}) = 1.2 \cdot 10^{-11}$, as would be the case with widespread defects, we have

$$\bar{t} = \frac{LLR}{RS_B} \cdot 1.2 \tag{AII.1}$$

Here LLR is the design internal load for damage tolerance, and RS_B is the residual strength B-value; this is the total effect of safety for the requirements of elements of safety for a given number of processes among the elements of safety.

Appendix III

Potential Defects in Design Criticalities, Elements of Safety Processes, Purpose, Rationale and Strategy

The White House Commission on Aviation Safety and Security wrote the Final Report in 1997. The report concluded that the Transport category of aircraft needed improvement and that the regulatory and advisory materials of the future needed to include objective measures of safety and required safety improvement, so that the safety levels and improvements of the future would be "performance driven." It set the baseline safety for the transport category a "One fatal crash out of 10^6 flights" and that "a five-fold safety improvement must be realized by 2007" and by implication most other categories were worse off.

The FAA Committed to a Safety Management System Principle in FAA order 8040.4 and with help from the GAO (Government Accounting Office), they settled for a system that was not risk based, but inspection based, with definitions "as you go." The results for the period 2002–2005 fell short of objectives, with one fatal accident in $5 \cdot 10^{-6}$ flights, but with a record of excessively late flights which was deemed symptomatic of a growing safety problem reported by the GAO. The outcome for 2006 for passenger and cargo airplanes measured against a target of 1.8 fatal accidents per 10^7 flights. The accident records for cargo planes were still 2.5 times worse than the passenger airplanes, and the GAO found the 2006 performance unacceptable, because the number of operational errors was 4.21 per 10^6 flights (see late flights above). The year 2007 was predicted to miss the stated objectives mostly due to inadequate staffing.

The FAA have declared their intention to implement AC 120–92 (System Risk Management) in the future, making their targets and principles consistent with the recommendations with the final report from "The White House Commission on Aviation Safety and Security."

REGULATORY SUMMARY AND PURPOSE

The recommendations in the Final Report of the White House Commission of Aviation Safety and Security established the need for objective safety measures and measures of safety growth which establishes the need for basic, guiding equations for Safety Management.

The review by the GAO (2002–2007) of how well targets were met showed that the results for different category flight vehicles were vastly different.

The best performing category was passenger transports, which became the focus of the continued effort. The results for this category were rejected by GAO as unacceptable by the measures identified in the adopted final report. It should be recognized that the used statistical populations include very little composite, primary structure and still were rated unacceptable.

The future use of equations included would therefore be essential for both innovation and evaluation of existing models with substantial composite applications.

The purpose of this appendix is to produce the equations required to minimize the probabilities of Unsafe States, to be able to design varying composite structures to safety constraints, to eliminate practices that are detrimental to safety, and to incorporate solutions of "outside" problems like defects in ground-support.

Structural Safety Management is defined and discussed in the main part of this book. It deals with the interaction of the elements of safety and their influences on structural safety. There are five elements of structural safety discussed in the main body of the book and here. The safe states of these elements are:

1. Safe Design, S_D
2. Safe Manufacturing, S_M
3. Safe Maintenance, S_I
4. Safe Operation, S_O
5. Safe Requirements formulation, S_R

The equation defining total State of Safe Structure is:

$$P(S_T) = P(S_D S_M S_I S_O S_R)$$

And an Unsafe State of Structure obtained from the basic definition is:

$$P(S_E) = \overline{1} - (P\overline{S}_E)$$

Here the second term on the right-hand side of the above definition describes the probability of all random defects and aberrations of the unsafe state of a specific element.

The following equation represents one expansion of the equation for a total state of safe structure, where the order of the "given events" is irrelevant.

$$P(S_T) = P(S_D \mid S_M S_I S_O S_R) \cdot P(S_M \mid S_I S_O S_R) \cdot P(S_I \mid S_O S_R) \cdot P(S_O \mid S_R) \cdot P(S_R)$$

Substitution of the definition of an Unsafe State of Structure into the above equation gives us a picture of all events that contribute to an Unsafe State, and the initial equation becomes

$$P(S_T) = [1 - P(\bar{S}_D \mid S_M S_I S_O S_R)] \cdot [1 - P(\bar{S}_M \mid S_I S_O S_R)] \cdot [1 - P(\bar{S}_I \mid S_O S_R)]$$
$$\cdot [1 - P(\bar{S}_O \mid S_R)] \cdot [1 - P(\bar{S}_R)]$$
$$= 1 - [P(\bar{S}_D \mid S_M S_I S_O S_R) + P(\bar{S}_M \mid S_I S_O S_R)$$
$$+ P(\bar{S}_I \mid S_O S_R) + P(\bar{S}_O \mid S_R) + P(\bar{S}_R)] + \Theta(p^2)$$

The final steps lead to,

$$\overline{1} - P(S_T) = P(\bar{S}_T) = P(\bar{S}_D \mid S_M S_I S_O S_R) + P(\bar{S}_M \mid S_I S_O S_R) + P(\bar{S}_I \mid S_O S_R)$$
$$+ P(\bar{S}_O \mid S_R) + P(\bar{S}_R) - \Theta(p^2)$$

Here we have the Probability of an Unsafe State in terms of Probabilities of unsafe states of elements of safety. The round-off error, for this unsafe state with a requirement probability standard of less than 10^{-10}, is of the order of $<10^{-20}$ and consequently negligible. This provides the tools needed to fulfill the objectives of Safety Management.

The "Unsafe States" consists of a number of Elements of Safety events that define the defects associated with the designing, building, operating, and maintaining flying vehicles in a safe way, in this case passenger transports. The equation above defines a way to describe the detail unsafe states that make up the Total Unsafe State. The purpose of this analysis is to identify all the events that add up to the probability of a total Unsafe State and satisfy the safety requirements needed for designing, building, operating, and maintaining the vehicle safely.

The first term of this equation identifies events that affect the Total Unsafe State and is written as:

$$P(\bar{S}_D \mid S_M S_I S_O S_R) = P(\bar{S}_{DCM} \cup \bar{S}_{DPD} \cup \bar{S}_{DPE} \cup \bar{S}_{DPV} \cup \bar{S}_{DPC})$$

Here the following events are involved:

$\bar{S}_{DCM}$: Detrimental effect from Design Critical Mode

$\bar{S}_{DPD}$: Detrimental effect from Design Process Development

$\bar{S}_{DPE}$: Detrimental Process Execution

$\bar{S}_{DPV}$: Faulty Process Validation

$\bar{S}_{DPC}$: Failed Process Control

The second term of the focus equation can be expressed as:

$$P(\bar{S}_M \mid S_I S_O S_R) = P(\bar{S}_{MPD} \cup \bar{S}_{MPE} \cup \bar{S}_{MPV} \cup \bar{S}_{MPC})$$

Here the following events are included:

$$\bar{S}_{MPD} \;:\; \text{Detrimental effect due to Process Development}$$
$$\bar{S}_{MPE} \;:\; \text{Detrimental Process Execution}$$
$$\bar{S}_{MPV} \;:\; \text{Faulty Process Validation}$$
$$\bar{S}_{MPC} \;:\; \text{Failed Process Control}$$

The third term of the equation can be expressed as:

$$P(\bar{S}_I \mid S_O S_R) = P(\bar{S}_{IPD} \cup \bar{S}_{IPE} \cup \bar{S}_{IPV} \cup \bar{S}_{IPC})$$

Here the following events are participating in four activities: Scheduled Maintenance, Inspection, Reporting, Repair for the Maintenance Element (I) and they are defined as:

$$\bar{S}_{IPD} \;:\; \text{Detrimental effect due to Process Development}$$
$$\bar{S}_{IPE} \;:\; \text{Detrimental Process Execution}$$
$$\bar{S}_{IPV} \;:\; \text{Faulty Process Validation}$$
$$\bar{S}_{IPC} \;:\; \text{Failed Process Control}$$

The fourth term of the equation (complicated because of the need for "ground-support") is defined as:

$$P(\bar{S}_O \mid S_R) = \text{Typical processes } P(\bar{S}_{OPD} \cup \bar{S}_{OPE} \cup \bar{S}_{OPV} \cup \bar{S}_{OPC}) + \text{Emergency processes}$$
$$P(\bar{S}_{EOPD} \cup \bar{S}_{EOPE} \cup \bar{S}_{EOPV} \cup \bar{S}_{EOPC}) + \text{"Local Weather Reporting" Processes}$$

(Which for "wind-shear" is served by advanced technology, but has only limited application from a global perspective and "running out of Runway" is not an uncommon problem) requiring the following events be given attention in the "Local Weather Reporting" Processes

$$P(\bar{S}_{OWPD} \bar{S}_{OWPE} \cup \bar{S}_{OWPV} \cup \bar{S}_{OWPC}) + \text{ground damage reporting}$$

$$P[(\bar{S}_{OPD} \cup \bar{S}_{OPE} \cup \bar{S}_{OPV} \cup \bar{S}_{OPC}) \mid \bar{X}_t D_5] + \text{operational errors caused by lacking}$$

ATC staffing $P([\bar{S}_{OPD} \cup \bar{S}_{OPE} \cup \bar{S}_{OPV} \cup \bar{S}_{OPC}] \mid \bar{S}_{OAS}) = P(\bar{S}_{GPC} \mid \bar{S}_{OAS}) \cdot P(\bar{S}_{OAS})$

where $\bar{S}_{GPC}$ is the general process defect contribution and $\bar{S}_{OAS}$ is the defective ATC staffing.

The fifth term of the equation can be expressed as:

$$P(\bar{S}_R) = P(\bar{S}_{RPD}) \cup P(\bar{S}_{RPE}) \cup P(\bar{S}_{RPV}) \cup P(\bar{S}_{RPC})$$

Here the following are defect events associated with the Requirements Formulation Process:

$\bar{S}_{RPD}$: Detrimental effect due to Process Development

$\bar{S}_{RPE}$: Detrimental effects due Process Execution

$\bar{S}_{RPV}$: Faulty Process Validation

$\bar{S}_{RPC}$: Failed Process Control

The purposes of these equations are to evaluate and assess the probable defects and aberrations affecting different new and existing combinations of material types, manufacturing and structural concepts. Example 17.1 represents a coordinated illustration of DDP defect detection and prevention defects probabilities.

The probability of safe state is defined as the complement of the probability of an unsafe state. We now assume that the DDP-based probability can be considered as a representation of the 777 composites structure. We will present a numerical illustration before proceeding with the context.

The next equation has a very important role in providing a total picture of critical structural modes and Unsafe States.

The relation between the probability of an Unsafe State and the probability of Failure after *n* flights with an Undetected Loss of Limit Integrity (Unsafe State):

$$P(\bar{A}_t) = P(\bar{H}_\tau \bar{U}_t \bar{H}_t \bar{A}_n) = P(\bar{A}_n \mid \bar{H}_\tau \bar{U}_t \bar{H}_t) \cdot P(\bar{H}_\tau \bar{U}_t \mid \bar{H}_t) \cdot P(\bar{H}_t)$$

$$\rightarrow P(\bar{H}_\tau \bar{U} \bar{H}_T) \quad when \ n \rightarrow \infty$$

$$\therefore P(\bar{A}_t) \rightarrow P(\bar{S}_T) \quad when \ n \longrightarrow \infty$$

Different situations are shown below:

n flights	Probability of failure after *n* flights, $(p^{SU})^n$			$(p_f)^n$		
1	0.99	0.995	0.999	0.01	0.005	0.001
100	0.37	0.61	0.90	0.63	0.39	0.1
1000	0.00004	0.0007	0.37	1	1	0.63
2000	$1.9 \cdot 10^{-9}$	$4 \cdot 10^{-5}$	0.14	1	1	0.86
3000	$8 \cdot 10^{-14}$	$3 \cdot 10^{-7}$	0.05	1	1	0.95
6000	~0	$8.7 \cdot 10^{-14}$	0.003	1	1	0.997

The above table is an illustration of surviving n flights with undetected loss of Damage Tolerance integrity, and the last three columns is the probability of failure after n flights or less, given that the probability to fail in one flight.

The focus equation that defines the probability of an Unsafe State given by Safety Management with five Elements of Safety and raises the question of target probabilities for different kind of defects.

Example AIII.1   The first term of the equation is expressed as a sum of defect-events. The criticality of the first event is described above. The probability of an undetected loss of integrity in the right design can be expected to be 10^{-12}. The following process quality (P_Q) would lead to the long-term probabilities of defect-free zones Z_{DF}

P_Q	Z_{DF}
6.5σ	$4.2 \cdot 10^{-11}$
7.0σ	$1.3 \cdot 10^{-12}$
7.5σ	$3.2 \cdot 10^{-14}$

A review of 17.1.3 would yield the total contribution. The nature of the example material yields a contribution from the critical mode of 10^{-12}. The assessment of the contribution from the design process development of the auxiliary contribution from the other types of defects is dealt with as initially fractional entities and let the iterative totalities be adjusted as required probabilities accumulate. The contributions from the first term of the focus equation add up to $10^{-12}+(1+.1+.1+.1) \cdot P(\overline{S}_{PD}) = 10^{-12} + 1.3 \cdot P(\overline{S}_{PD})$. The table above shows a few possible alternatives for well-defined defects-free zones for consideration. The requirements will be selected from the table and the feasibility of the 10% contributions will be assessed in the process.

The contribution of the second term of the focus equation will become

$$n_m \cdot 1.3 P(\overline{S}_{PD})$$

by analogous arguments to above, and n_m being the number of manufacturing processes employed at a DDP.

The contributions by the third term analogously become:

$$n_i \cdot 1.3 \cdot P(\overline{S}_{PD})$$

where n_i is the number of processes active in Maintenance at the DDPs.

The contributions by the fourth term are more complicated for processes active in Operation at the DDPs.

For Standard Processes the contributions are:

$$n_{so} \cdot 1.3 \cdot P(\overline{S}_{SPD})$$

where n_{so} is the number of standard processes.

For Emergency Processes the Contributions are:

$$n_{eop} \cdot 1.3 \cdot P(\overline{S}_{EPD})$$

where n_{eop} is the number of Operation Emergency Processes.

For Weather Reporting Processes (including Wind-shear, General Landing Wind State, Hail-storms in flight path), contributions are:

$$3 \cdot 1.3 \cdot P(\overline{S}_{WRP})$$

For Ground Damage Reporting Processes contributions are:

$$n_{gdp} \cdot 1.3 \cdot P(\overline{S}_{GDP})$$

where n_{gdp} is the number of processes (depending on damage location).

For the Ground Support Quality depending on ATC-staffing, the probability contribution is:

$$n_{acp} \cdot 1.3 \cdot P(\overline{S}_{ACP} \mid \overline{S}_{OAS}) \cdot P(\overline{S}_{OAS})$$

where $\overline{S}_{OAS}$ is the defective ATC staffing, and n_{acp} is the number of affected processes.

The sum of the above contributions is $P(\overline{S}_{OP})$.

Finally the fifth term amounts to this contribution:

$$n_{rp} \cdot 1.3 \cdot P(\overline{S}_{RP})$$

where n_{rp} represents the number of requirements processes.

A desirable strategy is to assign the same quality to all the defects process zones, assess how it compares to the overall requirement limits from an attainable and practical standpoint.

The Structural Safety Management discussed here is therefore based on objective targets for the Probability of Unsafe States complemented with Risk management and

Uncertainty reduction. Section 17.1.1 describes the tie-in between probability of an unsafe state and the probability of failure.

The probability requirements for the following are:

The airplane (one fatal accident out of 10^6 flights)	10^{-6}
The Structure is assigned 10% of the total	10^{-7}
The number of PSEs is N_P	$10^{-7}/N_P$
The average number of DDPs is N_D	$10^{-7}/(N_P \cdot N_D)$
Target Safety increase & $N_P = 100$ & $N_D = 30$	$33 \cdot 10^{-10}$

It is clear, that if instead, the number of PSEs is equal to 100 and the number of DDPs is 30, a somewhat different situation emerges, and there is a need to analyze the drivers for safety on the DDP level. In this case however, with a three-fold safety improvement as recommended by The White House Commission in the Final Report, the objective becomes 10^{-11}, – a reasonable first objective, considering the expectations for composites.

The basic strategy used here, which would lead to a requirement of 10^{-11}, would lead to a process quality requirement in the vicinity of 10^{-14} could put serious strain on the process development, and could be helped significantly if the structural concepts were designed Fail-Safe.

The basic aspects of the Elements of Safety and the way the basic safety requirements translate to structural design requirements depend on a resolution of competing safety needs. This in turn has to translate to safety models for structural safety, which as we have seen in previous chapters, combine the descriptions of models for structural, internal loads, the needs of manufacturing, maintenance, operations and requirement formulation. The basic safety model has to start at the DDP level in general and the PSE level for Discrete Source Damage Tolerance and communications "Bottoms-up" and "Top-down" must be possible.

The Elements of Safety and potentially critical modes in the realm of Discrete Source Damage Criticality have the potential to be critical in both Damage Resistance and Damage Tolerance. Until Extensive testing of different discrete source damage designs – including the point design capabilities, the next step into design to safety constraints must have a firm testing foundation. There are no analytical damage resistance and damage tolerance methods and Point design approach is presently the only way to produce PSE-based design for many parts of the airplane. Recognizing that new materials, new manufacturing and structural concepts result in new criticalities and require extensive new testing will be the creative way to innovate the world of composites.

Bibliography

Backman, B.F. (2005) *Composite Structures, Design, Safety and Innovation*. Elsevier, Amsterdam and London.

Chatfield, C. (1988) *The Analysis of Time Series*. Chapman and Hall, London and New York.

Congdon, P. (2001) *Bayesian Statistical Modelling*. John Wiley & Sons Ltd, Chichester.

Feller, W. (1957) *An Introduction to Probability Theory and Its Applications*, 2nd edn. John Wiley & Sons Ltd, New York.

Martz, H.F. and Waller, R.A. (1982) *Bayesian Reliability Analysis*. John Wiley & Sons Ltd, New York.

Sheldon, R.P. (1982) *Composite Polymeric Materials*. Applied Science Publishers, London and New York.

Shooman, M.L. (1968) *Probabilistic Reliability, An Engineering Approach*. McGraw-Hill Book Company, New York.

Tribus, M. (1969) *Rational Descriptions, Decisions and Designs*. Pergamon Press Inc., New York.

Ward, I.M. (1983) *Mechanical Properties of Solid Polymers*, 2nd edn. John Wiley & Sons Ltd, Chichester and New York.

Index

 Index